TAM

Marine Populations

Marine Populations
An Essay on Population Regulation and Speciation

Michael Sinclair

BOOKS IN RECRUITMENT FISHERY
OCEANOGRAPHY

Washington Sea Grant Program
Distributed by University of Washington Press
Seattle and London

Publication of this book is supported by grant NA86AA-D-SG044 from the National Oceanic and Atmospheric Administration to the Washington Sea Grant Program, projects E/FO-2 and A/PC-5. The U.S. Government is authorized to produce and distribute reprints for governmental purposes notwithstanding any copyright notation that may appear hereon.

Cover illustration (paperback) by Fred Lisaius

Library of Congress Cataloging-in-Publication Data

Sinclair, Michael, 1944–
    Marine Populations: an essay on population regulation
and speciation / Michael Sinclair
       p. cm—(Books in recruitment fishery
oceanography)
      Bibliography: p.
      Includes index.
      ISBN 0–295–96633–5. ISBN 0–295–96634–3 (pbk.)
      1. Marine ecology. 2. Population biology.
      3. Speciation
    I. Title. II. Series.
QH541.5.S3S49 1987
574.5'2636—dc 19
                         87–37116

CONTENTS

FOREWORD

Recruitment fishery oceanography can be defined as the study of the effects of environmental variability on year class strength, or recruitment, in populations of marine organisms, especially those of commercial importance. Such studies are concerned with the factors that determine the continuing productivity of living resources under environmental and fishing stress. As such they have important implications for fishery management. They are also of great scientific interest. For example, the biotic consequences in the ocean of large-scale climatic change ("global change") are probably best revealed in population trends of commercial fish stocks.

Studies of recruitment fishery oceanography are inherently interdisciplinary, involving as they must knowledge of meteorology, fishery science, and physical, chemical, and biological oceanography. Many of the elements of the mechanics of interaction are available, but only recently have there been serious attempts to assemble them in a rational way.

To accelerate work on the synthesis of ideas, as well as to communicate current thought to scientists at the University of Washington and elsewhere, the Sea Grant Seminar Series on Recruitment Fishery Oceanography was established in 1985, with Dr. Michael Sinclair delivering the first lectures in December of that year. His lectures are published here as the first volume in the Washington Sea Grant monograph series, Books in Recruitment Fishery Oceanography.

Warren S. Wooster
Institute for Marine Studies
University of Washington

BOOKS IN RECRUITMENT FISHERY OCEANOGRAPHY

Editorial Staff

Patricia Peyton
Manager

Alma Johnson
Editor

Margaret Darland
Production Editor

Victoria Loe
Designer

Christine Beecham
Typesetter

Faculty

Karl Banse
School of Oceanography
University of Washington

Robert C. Francis
Fisheries Research Institute
University of Washington

Warren S. Wooster
Institute for Marine Studies
University of Washington

Preface

The stimulation to present at this particular time some thoughts on population regulation from a general rather than a restricted scope was generated by an invitation from Warren S. Wooster, of the Institute of Marine Studies at the University of Washington, to give a series of lectures on the population biology of Atlantic herring. Given the academic setting for the lectures, I decided to broaden the scope beyond the stock assessment framework within which much of my research has been conducted and to consider the marine fisheries population biology literature in relation to the more fundamental ecological literature. In essence, the aim has been to evaluate the relevance of the stock assessment findings on commercially exploited marine species to the more general question of the regulation of animal abundance. Much of the joint work done with colleagues on herring, which is already in the primary literature, has been summarized. This has been done both because the request for the lectures was based on the herring papers and because the generalizations drawn for Atlantic herring form the basis of the concepts on population regulation that are further developed here. This essay predominantly advocates a particular point of view and thus may be criticized for giving less attention to alternative models.

Most of the ideas in this essay are the result of joint work with my colleague and friend, T. Derrick Iles. His generous exchange of ideas and insights has been critical to the development of my present interests in population biology. His contribution, due to our collaborative ongoing research, is fully and gratefully acknowledged. Sharon LeBlanc contributed enormously by rapid and accurate treatment of the onerous job of processing and editing a growing and continuously changing manuscript. The library staff of the Halifax Fisheries Research Laboratory, in particular Lynne Cook, were most helpful in arranging loans of books and papers that are not available locally. It has been a delight working with both Alma Johnson, the editor, and Patricia Peyton, the communications manager of the Washington Sea

Grant Program. The editorial contribution is gratefully acknowledged. The illustrations were drawn by Gail Jeffery, and translations of parts of several key references in German were written by Elizabeth Hill. Earlier drafts of the manuscript were critically reviewed by K. Banse, J. F. Caddy, M. Dadswell, R. W. Elner, A. Fleminger, T. D. Iles, S. Kaartvedt, S. Kerr, A. Koslow, A. Longhurst, A. D. MacCall, P. Pepin, R. Stephenson, R. Strathmann, M. J. Tremblay, and S. Wilson. I thank them for their help and note that there are substantive points of disagreement in some subject areas.

This essay is dedicated to the marine fisheries researchers of the first fifty years of the development of this discipline (from about 1880 to 1930), who laid such a fine groundwork for the subsequent conservation and management of fisheries—in particular D. Damas (a Frenchman), T. W. Fulton (a Scot), F. Heincke (a German), J. Hjort (a Norwegian), and J. Schmidt (a Dane). Much of their research was done within the framework of the International Council for the Exploration of the Sea (ICES). There is a tendency from a near perspective to lose sight of the substantive increases in understanding of population biology that have been generated by practical stock assessment research and to emphasize the shortcomings of management initiatives. From a historical perspective, however, the contribution of fisheries research to both rational management and marine ecology is most impressive.

Finally, I wish to thank W. S. Wooster, the past president of ICES, for his invitation to present, and subsequently publish, this essay based on the series of lectures. The text follows closely the lectures which were presented in December 1985.

···1···

THE POPULATION REGULATION QUESTION

In some ecological studies, the term *population* simply means all those animals or plants of a particular species within a chosen study area (such as a population of barnacles, *Balanus balanoides*, colonizing rock surfaces on the shore in the Firth of Clyde, Scotland, as studied by Connell [1961]). In such studies the population under examination is not assumed to be a self-reproducing unit. The population may be an abstraction rather than a natural unit, i.e., an abstraction of the ecologist for the particular questions being posed.

In other field studies the same term, sometimes lengthened to *local population*, defines a different concept, that of a self-sustaining component of a particular species. The geographical distribution of the population in this latter sense can be defined without reference to an ecological question. To the degree that the population is accurately delineated, it is clearly a real phenomenon of nature rather than an ecologist's abstraction.

Throughout this essay it is the latter concept of a population that is explored. In sexually reproducing animals (which are the focus of this perspective), the abundance of individuals in a population represents the size of the shared gene pool (the effective population size in population genetics). Links from ecological studies to systematics and evolutionary studies are facilitated when the second definition is employed. These are the populations implicit in the biological species concept—"groups of actually or potentially interbreeding natural populations, which are reproductively isolated from other such groups" (Mayr 1942, p. 120)—and the *Rassenkreis* of Rensch (1929).

The characteristic of absolute abundance of animal populations has received surprisingly short shrift in the ecological literature even though an understanding of abundance is critical to flanking disciplines such as population genetics and evolutionary biology. Dobzhansky

(1937), using *population* as adopted in this essay, states (p. 138), "It is no exaggeration to say that the conclusions which eventually may be reached on the dynamics of the evolutionary process will depend in no small degree on the information bearing on the problem of the population numbers." Notwithstanding a rich literature on population regulation and its central role in mathematical population genetics and evolutionary theory, an understanding of the control of abundance has not materialized. Slobodkin (1972), in a stimulating paper commemorating G. E. Hutchinson's research contribution, argues that abundance, like heat, is a globally extensive variable that, unlike temperature, cannot be measured by the individuals within a population. Thus, abundance, unlike density (the number of individuals per unit area or volume), cannot be internally regulated.

The study of the control of local *density* of individuals was a major focus within animal ecology from the 1930s to the 1960s. There were two well-defined camps and the debate between them was heated, even acrimonious at times. Their fundamental disagreement involved the importance of the inverse effect of density in determining the amount and direction of change in the numbers of organisms in a given study area (Slobodkin et al. 1967). The central issue was whether density-independent factors alone could control the average numbers and the variance observed. The debate dissipated without settlement in the early 1970s and is no longer a topical area in ecology. Stearns's (1982) assessment, in a review of progress in evolutionary ecology in the twenty-five years after 1956, is perhaps generally accepted (p. 635):

> The most controversial outstanding ecological problem of 1956 was whether natural populations were controlled primarily by abiotic factors like temperature and moisture that operated independently of the density of the affected population, or whether they were regulated primarily by biotic factors like competition and predation whose impact depended on density and generated negative feedback. This controversy subsided without any clear resolution . . . , in the sense that few would now be willing to argue that we understand what generally regulates natural populations.

In this essay we argue, as do Ehrlich and Birch (1967) and Ehrlich and Murphy (1981), that the failure to convincingly resolve the question of what controls density fluctuations was at least partially due to the lack of a satisfactory hypothesis accounting for the existence of geographical population patterns themselves. Why are some species composed of many populations (such as the checkerspot butterfly, *Euphydras editha*) and others of only a few (such as the satyrnine butterfly, *Erebia episodea*) (Birch 1970, pp. 117–118)? Why does the yellow-bellied sea snake, *Pelamis* sp., comprise a single broadly distributed population of high abundance (Kropach 1975), whereas the beaked sea snake (*Enhydrina schistosa*) is represented by numerous populations of low abundance (Voris 1985)? This characteristic of species (the number of populations) is called *population richness* in the essay.

The above examples of differences in numbers of populations within a species are chosen, not because they are exceptional, but because they are reported in recent ecological literature. The evidence is massive that population richness varies between species and that a geographic population pattern is a characteristic of most animal species (the earlier research results are summarized by Mayr [1942, chap. 6] and Rensch [1959, chap. 3]). The research was predominantly carried out by systematists and has not been sufficiently incorporated into the modern ecological literature. These exhaustive surveys of the complex population patterns within the overall distributional limits of a species helped to define the biological species concept. Also, the geographical distributional evidence from what Mayr (1982a, p. 1130) has called "population systematics" was a critical contribution to the evolutionary synthesis itself (Mayr 1982b). (The term "evolutionary synthesis" is used as Mayr and Provine [1980] use it, to identify the synthetic literature on evolutionary biology from the 1930s to the 1950s. The other term in common usage to identify this body of literature is "modern synthesis.")

Paradoxically, even though the patterns of populations were exhaustively described by systematists, there has been no general theory developed by ecologists to account for species-specific differences in population richness. Without a plausible interpretation of the processes generating the patterns in populations it may have been premature to attempt to resolve the question of the control of absolute abundance of a particular population. Essentially, the research question may not have been sufficiently split into its component parts to permit successful resolution.

Four components of the general question of population regulation can usefully be defined:

1. What determines the differences between species in population richness?

2. Why are the component populations of a particular species distributed in the observed geographical patterns?

3. What processes control the absolute abundance of the individual populations (or what controls their means)?

4. What processes control the temporal fluctuations in abundance of the individual populations (or what controls their variances)?

Almost all of the debate in the 1950s and 1960s focused on the final question in isolation from the first three questions, and it addressed the variance in density rather than in abundance.

The four aspects of population regulation are schematically

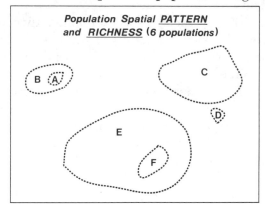

Fig. 1.1 Schematic representation of the four components of the population regulation question: pattern, richness, abundance, and variability.

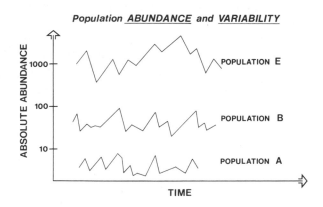

represented in Figure 1.1. Note in the upper part of the figure that small populations A and F are within the distributional area of populations B and E. In other words, reproductive isolation between populations occurs without geographic barriers. This represents a fairly common observation in marine fish species (Atlantic cod and Atlantic herring, for example).

This essay addresses the four components of population regulation in the oceans, with particular emphasis on marine fish. Chapter 2 presents a historical sketch of the development of concepts. The sketch is useful to the extent that it can demonstrate why advances in understanding of population regulation in general may be derived from the present marine fisheries and biological oceanographic literature. Chapter 3 reviews some recent developments on the population biology of Atlantic herring in detail. To a certain degree the concepts developed for Atlantic herring are generalized for the marine environment as a whole in subsequent chapters. The summary of the herring work (which includes a critique of the extant hypothesis of recruitment variability for marine fish) concludes the background material, or the setting.

The member/vagrant hypothesis is then stated in Chapter 4 to account for the four aspects of the regulation of sexually reproducing populations. Sexual reproduction itself is considered to generate critical constraints. The hypothesis emphasizes the importance of spatial constraints in the oceans to life-cycle closure. For species with complex life histories, populations are considered to be regulated, to a large degree, by physical oceanographic processes. It is also concluded that competition for food resources, and food-chain interactions, is not necessarily critical to the regulation of abundance. However, for species with internalized or brood-carried early life-history stages (i.e., species with "collapsed" life histories), food-chain processes are considered to be much more important.

In three subsequent chapters (5, 6, and 7), the marine literature is reviewed to provide support for the member/vagrant hypothesis (including aspects of the fisheries, estuarine, oceanic island, zooplankton, and benthos literature). The evidence supporting the hypothesis is then summarized in Chapter 8. In Chapter 9, assuming that the hypothesis has some support, implications for selected ecological issues are discussed. That discussion completes the second component of the essay.

Up to this stage, the treatment of population regulation, while unorthodox (in that pattern, richness, and abundance are highlighted rather than variability), is strongly based on empirical observations in the oceans. The third component of the essay (Chapters 10, 11, and 12) is considerably more speculative, and may grate on the reader because of possible unsophisticated use of terms or lack of appropriate balance. This is perhaps unavoidable given the breadth of the subject matter that is addressed. In spite of this difficulty, the implications of the member/vagrant hypothesis for flanking disciplines are worth a tentative exploration. In the second component of the essay it is concluded that for sexually reproducing marine species *with* complex life histories, population regulation does not necessarily require competition for limited resources or food-chain interactions for stability. If robust, what are the implications for the ecological aspects of evolutionary theory? Again, a historical sketch (Chapter 10) provides perspective for the reader unfamiliar with conceptual developments in evolutionary biology. The concepts of life-cycle selection (which includes sexual and mate selection) and food-chain selection are then developed as natural extensions of the two classes of losses of individuals identified in the member/vagrant hypothesis. It is further argued that life-cycle selection plays the predominant selective role in speciation (the origin of reproductive isolation), with food-chain selection involved in so-called energetics adaptation. In sum, adaptation and speciation are considered to be separate processes associated with identifiable components of population regulation.

Although the shifts in emphasis implied by the member/vagrant hypothesis are modest, their implications may be substantive because the shifts are at the conceptual core of the ecological mechanism of evolution (i.e., population regulation). For example, empirically observed population richness in the oceans is not necessarily evidence of speciation in action. Location in appropriate space to contribute to population persistence, rather than increases in numbers, is argued to be central to the selection process for sexually reproducing animals. Geographic opportunities permitting life-cycle closure, rather than geographic barriers, are argued to be key aspects of both population regulation and speciation. Finally, much of the confusion of the density-dependent versus density-independent debate on population regulation is argued to be at least partially due to inappropriate biological units of study. Population thinking in the sense of Mayr has been absent to a large degree in modern ecology (Kingsland 1985).

···2···

HISTORICAL BACKGROUND OF
MARINE POPULATION CONCEPTS

The study of population regulation in the oceans-began in the late 1800s. Several major conceptual developments relative to the four questions already stated can be traced since that time. More than a century earlier, Buffon (1758, quoted in Cole 1957) inferred that biological interactions, including both competition and predation, were important in the regulation of marine fish abundance. Atlantic herring, he noted,

> present themselves in millions to our fishermen, and after having fed all the monsters of the northern seas, they contribute to the subsistence of all the nations in Europe for a certain part of the year. If prodigious numbers of them were not destroyed, what would be the effects of their prodigious multiplication? By them alone would the whole surface of the sea be covered. But their numbers would soon prove a nuisance; they would corrupt and destroy each other. For want of sufficient nourishment their fecundity would diminish, by contagion and famine they would be equally destroyed; the number of their own *species* [emphasis added] would not increase, but the number of those that feed upon them would be diminished.

Although his ideas on the species concept anticipated modern developments, Buffon nevertheless supported the concept of constant, well-developed species (the essentialist species concept) (Mayr 1982a, pp. 261–262); and thus component populations were not yet recognized. Studies treating the changing species concept, i.e., the replacement of the essentialist species concept by population thinking, were clearly critical to the development of population regulation concepts themselves (defined in this essay to include pattern, richness, abundance, and variability). Species had to be envisioned, not as types, but as groups of populations before ecological concepts of the regulation of population abundance could be developed.

Two somewhat separate research thrusts can be identified in the late nineteenth century literature in marine studies dealing with the changing species concept. First, applied fisheries scientists focused their research on the causes of interannual variability in abundance of commercially important finfish species as indicated by marked differences in annual landings and catch rates (catch per unit of effort). Second, academic scientists addressed the more fundamental question of intraspecific variability in body form in relation to topical evolutionary questions generated by Darwinism. The marriage of the two thrusts led rapidly to what can legitimately be called a paradigm shift in fisheries biology at the turn of the century. The key contributions to the population question (i.e., the very existence of populations in the oceans) were made by F. Heincke, of Kiel University and director of the Biological Institute of Heligoland, and J. Hjort, director of the Board of Sea Fisheries at Bergen and later a professor of marine biology at the University of Oslo.

Heincke's contribution (summarized in his 1898 paper) was twofold. He established the methodology for the identification of fish populations, and he exhaustively described discontinuous geographic variability of form in Atlantic herring (*Clupea harengus*). To do this he independently developed statistical concepts and sampling theory prior to the flowering of mathematical statistics in the United Kingdom led by K. Pearson. As a result of the methodology he developed, he was able to demonstrate convincingly that the Atlantic herring comprises a large number of self-sustaining populations. This was a hotly debated issue in marine research at the time. Its resolution by a statistically rigorous methodology had broader implications. Dimorphism in form as evidence of speciation in action by natural selection was actively being investigated by the so-called biometricians, particularly in the United Kingdom. Heincke's earlier findings on populations of herring (Heincke 1878, 1882) had an influence on the initial focus of the Evolution Committee of the Royal Society, which was formed in 1894 and chaired by F. Galton, with W. F. R. Weldon as secretary. The first study by the committee was a statistical analysis of form using a sample of Atlantic herring taken from the Plymouth spawning population (Pearson 1906). Between 1890 and 1905, mathematical statistics made great progress, stimulated by the need to develop new theory and techniques to address, from an evolutionary perspective, population questions of the type highlighted by Heincke for Atlantic herring.

Heincke's work on Atlantic herring, because it so convincingly undermined the species concept itself, rapidly became known throughout the research community both at fisheries laboratories and in the universities. Schmidt (1917, p. 325) states the importance of Heincke's work on the existence of populations within species:

> Local races [populations] have in course of time been shown to exist in quite a number of fish species. Most important of all, in my opinion, are Heincke's herring investigations, collected in the comprehensive "Naturgeschichte des Herings," 1898. From a *mere chaos* [emphasis added], Heincke succeeded, by his admirable and systematic work, in furnishing not only a basis for all future investigations in this field; he also succeeded, through the study of a single species, the herring, in revealing so many important features—quite unexpected in part—as to [the] occurrence and relationship of various races [populations], that subsequent investigations with other species have in a certain degree only amounted to a repetition of Heincke's results.

It is difficult after the fact to fully appreciate the impact that such observations on populations had on the work of systematists. Goldschmidt (1940), however, captures it well (p. 27):

> I remember distinctly the shock which it created in my own taxonomic surroundings (I was an ardent coleopterologist at that time) when Matschie claimed that the giraffes and other African mammals had many different subspecific forms characteristic for different regions which he could recognize with certainty; when Kobelt claimed that the mussel *Anodonta fluviatilis* was different in each river or brook; when Hofer stated that each Alpine lake contained a different race of the fish *Coregonus;* or when Heincke claimed the same for herring.

Such observations led to the new species concept: the *Rassenkreis* of Rensch in the early 1930s in Germany and the so-called biological species concept popularized by Mayr shortly afterwards. Goldschmidt (1940) indicates that Heincke's detailed work on herring predated the taxonomic reform by several decades (p. 31):

> I might mention one such case [the requirement for a statistical approach in defining populations] in order to show that a con-

ception very similar to the rassenkreis concept had been arrived at in a very different way prior to that taxonomic reform. The herring in the North Sea forms large schools which are found in definite localities and travel to definite spawning grounds. These localities are different over the whole area inhabited by the species, and each area has a different constant race which, however, cannot be distinguished by ordinary taxonomic methods. Only a biometric study of a series of variable characters like number of vertebrae, number of keeled scales, and about sixty others, and their evaluation by biometric methods, permitted Heincke (1897–98) to find the constant racial differences. Since that time similar work with identical results has been performed by many ichthyologists.

Hjort's contributions (brought together in his classic paper of 1914) generalized the findings of Heincke on herring to other commercial species (Atlantic cod [*Gadus morhua*] and haddock [*Melanogrammus aeglefinus*]) and clearly identified the significance of "population thinking" to fisheries management. These conceptual developments were made possible by a critical development in methodology. In order to elaborate on Heincke's work on herring populations, a more detailed multinational study on the distribution and population patterns in the North Sea was carried out immediately after the formation, in 1902, of the International Council for the Exploration of the Sea. While analyzing the samples, H. Broch of Hjort's laboratory noted that the scales of the herring could be used to determine age. Under Hjort's direction a full study of determining age from scales was subsequently carried out by K. Dahl (who noted that the usefulness of scales for aging herring had been independently discovered by F. A. Smitt in Sweden a decade earlier). Scales also proved useful for distinguishing populations (Figure 2.1). The ability to age fish, in conjunction with the population concept so forcefully introduced by Heincke, generated considerable explanatory power. It was Hjort who recognized the full significance of the application of this new method of aging.

Prior to these contributions, annual variability in landings of marine fish in northern Europe was accounted for by the "migration theory" (Hjort 1914, pp. 4–5). It was thought that fish of a given species constituted a single integrated group throughout the distributional range (e.g., for Atlantic cod, from south of Cape Cod, Massachusetts, across the northern Atlantic to the North Sea and Norwegian coast). The large interannual variability in catches of what were believed to be

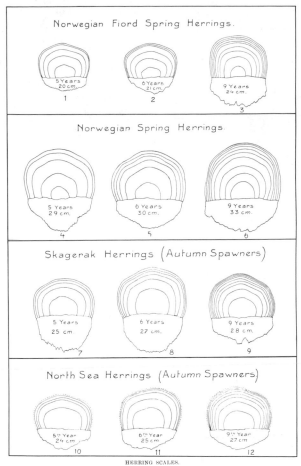

Fig. 2.1 Scales of Atlantic herring from different populations at the same ages (five, six, and nine years). (Reproduced from Dahl 1907)

rapidly growing fish species (life span of several years), as observed off the coasts of fishing nations, was thought to be caused by the interannual differences in oceanic-scale migration patterns of the species in question. However, with the aging technique it was possible to track the 1904 year class of a herring population through the Norwegian fishery over several years (Figure 2.2). It then became clear that the interannual variability observed in landings was due to year-class strength variability in age-structured populations. Hjort was able to generalize this observation by comparative analysis of herring, cod, and haddock.

In sum, the major contribution of Heincke and Hjort was to shift the focus of study in fisheries biology from the species to the

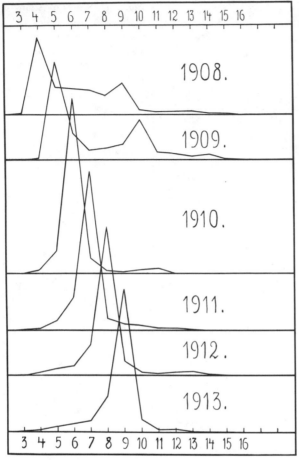

Fig. 2.2 Percentage age frequencies of samples from commercial landings of Atlantic herring in Norway from 1908 to 1913. The exceptionally strong 1904 year class dominates the landings throughout this period. (Reproduced from Hjort 1914)

population (in parallel with the more general shift in systematics from typology to population). In fact it appears that fisheries scientists as a whole incorporated the biological species concept into their disciplinary matrix well before the systematists themselves did; Mayr (1982a, p. 272, fn. 13) indicates that this occurred during the 1930s in systematics.

This paradigm shift provided great explanatory power; for example, variability in landings was then understood to be caused by interannual differences in recruitment to age-structured, geographically defined populations (of species with life spans of decades rather than a few years) rather than by the vagaries of panoceanic migrations of "typological" or "essentialist" species. The striking differences in the quality of fish landed (such as the proportion of cod liver oil in the landed weight) could then be readily explained as a result of the changes in mean age of a nonequilibrium population as year classes of different numerical abundance moved through the fishery.

In a more practical aspect, Hjort also established that the fishery itself could be used as a research tool if landings were representatively sampled at the dock. This simple contribution was critical to research on the population question, in that the spatial and temporal scales of sampling could match the actual scales of the biological phenomena being studied (hundreds of kilometers and years).

The shift toward population thinking generated intense study, still ongoing, on what causes the differences in year-class strength of marine fish populations. We shall return to this particular theme after first considering further developments relating to the question of population richness.

Heincke and Hjort established the population as the preferred unit of study for fisheries management questions. Subsequently, J. Schmidt, in a series of papers on "racial investigations" (1917 to 1930), made two other major contributions. He demonstrated experimentally that the differences observed between populations of marine fish species are at least in part genetically based, and he very clearly demonstrated that population richness varies dramatically between species. Four species received particular attention: Atlantic cod, Atlantic herring, European eel, and eelpout, *Zoarces viviparus*. By analyzing geographic differences in form, Schmidt concluded that population richness in these species ranges from one extreme, *Zoarces* sp., which has many populations, through cod and herring, to the European eel, for which no differences in meristics could be distinguished from samples throughout the rivers of Europe. He also recognized that there are differences in absolute abundance between populations (1930, p. 26): "The result of the investigation is that the North Atlantic cod, in contrast to the eel (*Anguilla vulgaris*) but like *Zoarces*, consists of a mosaic of populations dissimilar to one another. Some of these, particularly those belonging to the open seas, are distributed to a considerable extent, while others are far more local." He asked the "why" question later defined by Mayr (1961) and inferred that events during the early life-history stage are critical to interspecific differences in population richness. To our knowledge Schmidt's initial thoughts on this question were the only steps that had been taken to answer it in the ecological literature as a whole until recent years.

In sum, with reference to the four components of the population regulation question identified in Chapter 1, rapid progress was made between the late 1800s and 1930 on the first component (the very existence of marine populations as well as interspecific differences in population richness). Little conceptual progress has been made on this issue since Schmidt's contribution.

The second component, the reasons underlying the specific geographic patterns in population structure in the oceans, also was addressed aggressively in this earlier period, with T. Fulton, D. Damas, and again J. Schmidt making key contributions. Much of Damas's original work on this issue is incorporated in Hjort's 1914 classic with acknowledgment. Fulton (1889, 1897) addressed several fundamental questions on the location and timing of spawning as well as the duration of the egg, larval, and postlarval stages. He observed that spawning of diverse commercially important marine fish exploited in the North Sea by Scottish fishermen was remarkably localized and generally occurred in offshore waters (outside the three-mile territorial sea), yet adults outside the spawning period as well as juveniles were very broadly distributed. He asked (1897, p. 369) why they spawn precisely where they do: "It cannot be held that the selection of spawning places by fishes producing pelagic eggs is fortuitous, for adults with ripening reproductive organs move out from the inshore waters at the approach of the spawning season and may return to them after their spawn is shed."

Fulton (1889) had hypothesized that the selection of areas offshore for spawning was determined by "the physical conditions in relation to the safety of the floating eggs and larvae and their transport to the places most suitable for the welfare of the young fishes derived from them"; to test his larval drift hypothesis he carried out studies of the residual surface-layer circulation of the North Sea using drift bottles (1895, 1897, 1900). His work resulted in the first clear statement of the importance of larval drift in the definition of geographic patterns in spawning location. The research was conducted, however, within the constraint of the essentialist species concept without consideration of populations.

In two extraordinary studies, Damas (1909), under Hjort's research direction, and Schmidt (1909) described collectively the geography of all the gadoid species (i.e., the geographic locations of the different parts of the life history) in the waters from France to Iceland.

As stated above for Fulton's work, the research may have been constrained by the essentialist species concept of the time (which was not fully replaced until after Hjort's 1914 paper). The major contribution from these massive studies was to generalize the observations that Fulton derived for the northern North Sea. Damas (1909, p. 2) summarizes his findings relative to spawning location as follows:

> We will show how the spawning distribution is enormously more restricted than the total area of distribution, that the region where major spawning occurs is different for different species, that it is defined by hydrographic conditions.[1]

With respect to the early life history stages,

> The fundamental idea has been to follow the early life history stages (eggs, larvae and post-larvae) during their progressive dispersal under the influence of currents, their passive migration from the spawning areas out to the geographical limits of the distribution of the species.[2]

Thus, several elements were generalized on a very large scale: spawning locations were observed to be different between species, discontinuous within a species, and very restricted spatially relative to the distribution of the species; eggs and larvae were interpreted to be dispersed widely in the surface waters by the residual currents. Two constraints hindered explanation of the complicated distributional patterns: the species concept itself, and limited understanding of the physical oceanography of the distributional areas of the early life-history phases. The first constraint was eliminated by Hjort's (1914) synthesis, in which populations were identified as the appropriate unit of study. Thus, the complex distributional patterns of spawning were a function of the population structure of the species. The specific geographic patterns themselves, however, were not plausibly interpreted.

The second constraint, inadequate understanding of the physical oceanography, has continued to hinder understanding of what generates the population structure of a marine species. Scientists, using the prevailing concepts of residual circulation, hypothesized the early life-history phase (the eggs and larvae) to be entirely passive and adapted for dispersal of the species (pre-1914) or the population (post-1914), as well as for transport to the juvenile distributional area. The spatial scale of the information on surface-layer residual circulation (as shown in Figure 2.3) was of limited utility for the scale of the population processes, and there was little emphasis on shorter time

Fig. 2.3 Residual surface-layer circulation of the North Sea inferred from drift bottle releases and recoveries. (Redrawn from Fulton 1897)

scales of variability. Explanatory power relative to the pattern observed in spawning populations was limited.

Hjort (1926) and Schmidt (1930) hoped for a general theory of marine populations that would account for both the specific geographic patterns observed and the interspecific differences in population richness. Hjort (1926, p. 30) states: "An attempt has been made to determine the spawning areas of the principal fish, plaice, herring, cod, haddock, to define the spawning migrations, the nurseries where the young fish develop, etc. It is hoped it may in this way be possible to find *the general laws for the appearance of biological groups*" (emphasis added).

Within fisheries research laboratories very detailed work has continued to this day to describe the distributional patterns of fish populations (annual migrations as well as life-history migrations). However, even the most recent synthesis (Harden Jones 1968), which involved a modest conceptual advance beyond the above-mentioned studies, falls short of the desired "general law of biological groups" of Hjort. Harden Jones (1968) hypothesizes that life-history migrations, including egg and larval drift from spawning areas to juvenile nursery areas caused by surface-layer residual currents, generate the observed patterns of population structure (Figure 2.4). He assumes that eggs and larvae are distributed passively by the surface-layer residual currents from A to B in the figure. In essence the hypothesis incorporates the fundamental concepts of Damas, Fulton, Heincke, Hjort, and Schmidt as supported by the accumulated literature up to the late 1960s. However, as stated before, it is difficult to match population patterns of various species and their life-history distributions to the residual surface-layer circulations. In our view, Harden Jones's hypothesis involving the migration triangle has done little to explain population richness and pattern. This is discussed specifically for Atlantic herring in the following chapter.

The state of the art in physical oceanography at the turn of the century—and even in 1968, the date of the Harden Jones synthesis—critically restrained development of concepts on marine population regulation. It was perhaps the *wrong* physics for the time and space scales of distributional aspects of fish eggs and larvae. The circulation of the oceans has historically been derived from measuring the distribution of property fields (temperature, salinity, and density); direct measures of velocity are relatively rare. Using these scalar fields to generate an image of the residual circulation has stressed the mean flows to the detriment of the higher frequency variability. The information gained from the property fields should be viewed as a long-term average superimposed on considerable variability. The emphasis on a static view of the ocean circulation (i.e., time-independent motion) up until the 1960s is described by Wunsch (1981), who states (p. 344), "The question of the physical significance of a weak mean flow in the pres-

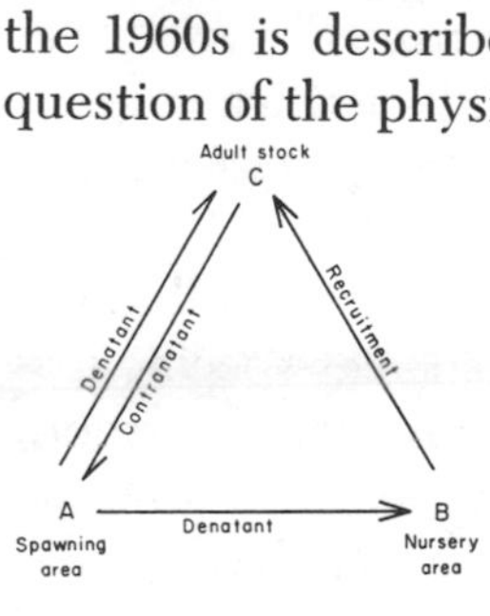

Fig. 2.4 Triangular pattern of fish migrations as generalized by Harden Jones (1968, Fig. 1).

ence of strong variability has rarely been addressed even now." The fisheries biologists still envision the ocean circulation in terms of the "weak mean flow," whereas the "strong variability" may be so important to eggs and larvae that the residual can be ignored or is irrelevant. This issue is central to the concepts elaborated throughout this essay.

In sum, the fusion of fundamental research on the species concept in the oceans (by Heincke and Schmidt, for example) with applied research on the causes of interannual variability in fish landings (as represented by Hjort, Fulton, and Damas) led both research and management to focus on populations. The implications of the population concept for fisheries management are as follows. Fishing in any part of the distributional range of a population influences the other parts of the range inhabited by the same population. Such fishing does not, however, influence the populations of the same species in contiguous areas. This straightforward yet powerful conclusion has led to the population as the desired unit for fisheries management. As a consequence, many descriptive studies have been undertaken on the geographic patterns of the populations and on the differences in population richness between species. Significant conceptual developments, however, have been sparse since the 1930s.

Many of the population patterns described by Heincke and the earlier marine population systematists have persisted in precisely the same geographical locations for more than a century. Also, populations that have recovered from commercial extinction have recovered in exactly the same geographical locations as they occupied before their collapse. One must conclude that the patterns in population structure are not transitory, and that certain features of the geography itself are critical to the persistence of the specific spatial patterns. At the edges of the distributional limits of a species and at biogeographical boundaries such as the Celtic Sea, where the Russell Cycle has been described (see Southward 1980), temporal variability is not unexpectedly greater at the discontinuity itself either in populations (e.g., Plymouth herring as described by Cushing 1961) or in species composition. This variability, however, should not overshadow the observed persistence of the population patterns. The persistence of the spatial patterns, as well as the variability, should be interpretable. Although the species-specific spatial patterns in populations are well described for many northern Atlantic commercially exploited species, the processes generating the patterns are not well understood.

There is a paradox in the development of the population concept in the oceans for zooplankton and for fish, what we call the "Damas paradox." Fish obviously have the ability to control their distribution against the ocean circulation during the juvenile and adult phases of their life history. Fish larvae, however, are almost universally depicted as passive drifters within the surface-layer circulation. Zooplankton complete their life cycle as plankton. Given the view of ocean circulation that prevailed in the early 1900s, it was difficult to account for the observed persistence in geographical space of zooplankton distributions. Damas (1905, 1909) and Damas and Koefoed (1907), working with both fish and zooplankton, attributed no behavior to ichthyoplankton by which their geographic distribution can be controlled (no doubt since it was felt that distributions could be defined subsequent to metamorphosis); on the other hand, Damas recognized that zooplankton have both daily and ontogenetic differences in behavior that cause them to persist in geographic space within diffusive and advective environments. This double standard between zooplankton and ichthyoplankton behavior (the Damas paradox) is still held, and will be discussed further subsequent to the statement of the new hypothesis on population in the oceans.

The development of concepts on the control of mean abundance and its variance (questions 3 and 4 stated in Chapter 1) has been hampered by the fact that theoretical studies and descriptive studies address the same question yet are frequently unrelated. The theoretical studies start with the initial premise that there is density-dependent control of abundance, and then develop equations to describe such a dependence between population size·and the recruitment generated (e.g., Ricker 1954; Beverton and Holt 1957; Cushing and Harris 1973; Ware 1980). Shepherd (1982), for example, states (p. 67):

> A versatile functional form relating recruitment to spawning stock biomass for fisheries is proposed. The non-negative, three-parameter form distinguishes resilience and degree of compensation as two aspects of the density-dependence of recruitment, and permits the representation of non-asymptotic, asymptotic, and domed stock-recruitment relationships by parameter variation within the same functional form.

In general, the fitted curves account for a small portion of the variance observed (see Figure 2.5), and it is frequently not possible to select

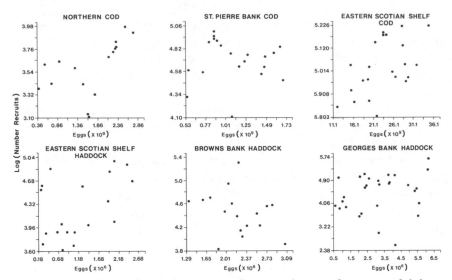

Fig. 2.5 Estimations of year-class size as a function of egg production or adult biomass for several populations of Atlantic cod and haddock in the northwestern Atlantic. (Redrawn from Koslow et al. 1987)

which of the theoretical relationships is preferred on the basis of the empirical results themselves. In addition, and perhaps more critically, the theoretical studies have not dealt with scale of abundance and the question of its control. Given that (1) absolute abundance is not addressed, (2) a very small part of the variance in year-class size is accounted for in most cases, and (3) "preferred" theoretical relationships cannot be determined on the basis of empirical observations, it is not inappropriate to conclude that the approach itself does not contribute to a resolution of the research question.

What is almost always observed under moderate fishing intensity is that fish populations have a characteristic abundance scale (egg production or adult biomass fluctuates over time within an order of, magnitude), but that year-class size fluctuates widely without obvious relationship to adult population size. The obvious question, then, is: Why is there such minimal relationship between population size and the recruitment generated (other than that the mature population size is within a characteristic order of magnitude for the population in question)? The mathematical treatments of this research question with fish, starting with Ricker (1954), evolved out of the emerging concepts of theoretical ecology, which emphasized the key role of competition and predator/prey interactions. However, theoretical studies have done little to increase our understanding of the control of either the mean or the variance in abundance.

The descriptive studies on the control of abundance in marine fish populations, in contrast to the theoretical studies, have dealt with the variance, but again not with the control of absolute abundance. As indicated above, Hjort made the first significant contribution to the variance question. In his synthesis (1914) he makes two important generalizations: (1) year-class size is determined early in the life history prior to recruitment to the fishery, and (2) year-class size is not a simple function of egg production. He discusses multiple causes of the inter-annual differences in mortality during the early life-history stages, yet he clearly states two hypotheses to account for them. The first hypothesis stresses the timing between spawning and phytoplankton blooms, both of which are variable. The cause of differential mortality between years is a result of differential food availability (young stages of zooplankton) at a critical stage in early fish larval development. The second hypothesis stresses the influence of drift of eggs and larvae away from the *appropriate* distributional area due to interannual differences in the circulation. The sources of mortality per se are not specified; instead, losses from the appropriate geographical areas for the populations are stressed.

Subsequent developments by the "descriptive" camp have predominantly elaborated on Hjort's first hypothesis. Conceptual advances have been modest. Cushing (1975) has formalized Hjort's "critical period" hypothesis in his attractively phrased match/mismatch theory. What he adds is a modified critical depth formulation that describes the interannual differences in the timing of phytoplankton blooms due to differential development of stratification/destratification of the water column in the spring/autumn (Gran and Braarud 1935; Riley 1946; Sverdrup 1953). The mechanism of larval mortality is again food limitation.

In essence, Cushing's match/mismatch theory hypothesizes that fish spawn in relation to the particular timing of spring and autumn phytoplankton blooms in the geographic area of inferred larval drift. In his hypothesis, fixed time of spawning coupled with a variable time of phytoplankton blooms, due to water-column stratification differences between years, generates variable fish larval survival. The theory is conceptually attractive, but support from field observations has been weak.

Several interesting wrinkles on the food limitation hypothesis have been added by Jones (1973), Laurence (1977), and Lasker (1975, 1978); but these contributions can be considered to support the match/mismatch theory at least in its frequently used looser sense (i.e., a match between larval development and larval food requirements rather than, as specified by Cushing, a match between timing of spawning and phytoplankton blooms). The steps in the development of the match/mismatch theory are summarized by Sinclair and Tremblay (1984). In sum, Hjort's food limitation hypothesis as elaborated by Cushing is perhaps still considered the extant hypothesis accounting for the control of population numbers of marine fish. There is a growing dissatisfaction with it, however; some recent tests of the match/mismatch theory are discussed in Chapter 3.

Since 1914, many studies have been published in support of Hjort's second hypothesis, that mortality or losses are caused by advection and diffusion of larvae out of the appropriate distributional area; the investigations by Sund (1924), Carruthers et al. (1951), Colton and Temple (1961), Iles (1973), and Johnson et al. (1984) are representative. These results, perhaps surprisingly, have not led to the generalization that spatial or geographic constraints are a major factor in the regulation of population numbers in the ocean.

The importance of population thinking in the sense of Mayr (1982a) was incorporated very rapidly into the disciplinary framework of fisheries biologists. Because of the critical importance of the definition of fish populations as fisheries management units, studies on the geographic population structure of marine fish species continued with vigor after the 1930s (at which time "population systematics" became old-fashioned in terrestrial ecology and in systematics itself); and it is still a high-priority research orientation in fisheries biology. As a result, there is a large literature (much of it of the gray variety) on geographic population structure of commercially important marine species. The literature cannot be introduced in a sketch of this length, and it is not well reviewed or synthesized.

There are a large number of observations on interannual variability in both adult population size and the year-class sizes generated. The ongoing port sampling of the landings of fishing fleets, as advocated by Hjort (1914), has led to the accumulation of observations on marine populations on the appropriate time and space scales for the biological phenomena of interest. This scale of population information is unique in the ecological literature. The applied insect literature has some parallels, but the terrestrial environment has been modified considerably more than the oceans and pesticides have been introduced. It is not inappropriate to conclude that population thinking in the sense defined in Chapter 1 has been developed more rigorously in fisheries biology than in any other field of biology.

By 1930 it was well recognized that population richness varies dramatically between species and, further, that events at the early life-history stages are important in generating the observed differences. The geographic patterns of populations for many species have been observed to persist over many decades. Although well described, a theoretical framework to account for pattern and richness in marine fish populations is lacking. In addition, theoretical and conceptual models of stock recruitment and recruitment variability have not addressed the question of the control of absolute abundance, even though empirical observations on populations of the same species indicate marked differences in means. Finally, there have been relatively minor advances in understanding the processes generating variability around the mean since Hjort's classic paper in 1914. Modern ecology has not significantly contributed to the resolution of the four questions on population regulation.

The contribution of the marine fisheries literature on populations to biology, and in particular to evolutionary theory, was considerable until the 1930s. Heincke's work on Atlantic herring, for example, is considered in some detail in Chetverikov's (1926) seminal contribution to evolutionary theory, and Schmidt's comparative studies on population richness were significant to the early development of ecologial genetics (Mayr 1982a, p. 553). Since the 1930s, however, the marine fisheries literature has contributed surprisingly little to theoretical developments in ecology and evolution. In Pianka's (1979) highly regarded text, *Evolutionary Ecology,* there is but one citation from the marine fisheries literature. More disturbing is that in Levinton's (1982) *Marine Ecology* fewer than two percent of the citations are from the fisheries literature. Margalef (1978, p. 662) in fact states, "Much of the work [in fisheries biology], published and unpublished, is uninspiring routine, although often wrapped in seemingly sophisticated jargon, and it is difficult to make use of much of the data that continue to flow into our libraries." This impression, which is perhaps generally shared, is unfortunate in our view because of the uniqueness of the information, in particular its characteristic time and space scales.

A general theory of marine populations, as hoped for by Hjort and Schmidt, has not materialized, but a considerable information base has accumulated following their lead. In spite of the rapid increase in understanding of richness and pattern (questions 1 and 2 in Chapter 1) between Heincke's first paper on herring variability in 1878 and Schmidt's last paper on racial investigations in 1930, few concepts have been developed since then on either of these questions or, for that matter, on the third and fourth questions. Fisheries biology made substantive contributions to flanking disciplines in what may be termed its "golden age" (about 1880–1930), but modern ecology has generated only modest increases in understanding of the population regulation question in the oceans. Other than Volterra's interest in fisheries problems, the accumulated research on the populations of marine fish during the golden age had essentially no influence on the development of animal population biology (see Kingsland 1985 and McIntosh 1985).

Population systematics, while critical to the evolutionary synthesis, is not the basis of population ecology. This may well be an important constraint in the development of ecology. The "population thinking" which was so important to the systematists has not been incorporated into ecological research.

···3···

POPULATION BIOLOGY OF ATLANTIC HERRING

A recent hypothesis on Atlantic herring (Iles and Sinclair 1982) deals with the broad scope of population regulation in the oceans (i.e., the four questions defined in Chapter 1). The initial questions it posed were:

- Why are there so many populations of Atlantic herring?

- Why do they differ in their characteristic levels of absolute abundance?

Secondary questions, concerning what controls the interannual variability in abundance, as well as the differences in spawning times between populations, were also addressed. This chapter summarizes the research on Atlantic herring and reviews critically the evidence in support of the match/mismatch theory.

Empirical observations indicate that there are many populations of herring, that the populations differ in mean abundance, and that they spawn at different times of the year. Since Heincke's research on herring between 1880 and 1900, considerable *indirect* evidence has accumulated which indicates that herring home to well-defined spawning locations to deposit their eggs on the sea floor (see Heincke 1898 and Harden Jones 1968 for summaries of early work).

That mature adults home to the spawning sites of their birth can be confirmed from the particular features of the Ile Verte herring population in the St. Lawrence estuary (see Auger and Powles 1980 and Henri et al. 1985 for distributional information). Herring of this spawning population can be identified from the unusual morphology of their otoliths, and they are so small at each age class compared with herring in the Gulf of St. Lawrence that they are frequently called "pygmy" herring. These pygmy herring spawn off Ile Verte in the estuary during the late spring, but some small herring are caught on the Magdalen Shallows in the Gulf of St. Lawrence during the feeding part of their annual migration. The larval phase of this population is exceptionally long (10–11 months; Sinclair and Tremblay 1984), in that the larvae from the spring spawn overwinter in this phase. The extremely slow larval growth in the upper St. Lawrence estuary is thought to cause the unusual morphology of the otolith. Thus, the link from the growth of the larvae within a very poor larval distributional area to homing of the adults to precisely the same geographic location to spawn is marked on the otolith. These observations, on an unusual population from a morphological perspective, provide *direct* support for the conclusion that Atlantic herring as a whole are philopatric (i.e., they return to reproduce at the site of their birth).

There is considerable evidence that absolute abundance varies between Atlantic herring populations. Iles and Sinclair (1982) and Blaxter (1985) independently selected representative examples from the literature of the range of population sizes (Table 3.1). Their studies agree closely. The range in abundances covers five orders of magni-

Table 3.1 Range in absolute size of herring populations.

Population	Spawning population size (*t*)*
From Iles and Sinclair 1982	
Blackwater estuary	100–500
St. Lawrence estuary	5,000–10,000
Iceland, summer spawners	100,000
autumn spawners	200,000
Southwestern Nova Scotia	400,000
Georges Bank	1,000,000
Norwegian	10,000,000
From Blaxter 1985	
Norwegian, 1950–66	2,500,000–10,000,000
North Sea, 1947–65	1,700,000–3,300,000
Georges Bank	272,000–1,140,000
Iceland, spring spawners	73,000–778,000
summer spawners	56,000–313,000
Baltic, 1966–67, autumn spawners	127,000–147,000
Celtic Sea, 1957–71	40,000–95,000
Elbe, spring spawners	Hundreds to thousands?
Blackwater, spring spawners	200–800
Lindaspollen	100

*Approximate population size during periods of moderate fishing.

tude. Under moderate fishing intensity the interannual variability in abundance for a given population is well within an order of magnitude.

The Herring Hypothesis

The empirical evidence, then, indicates that Atlantic herring populations have a wide range of characteristic absolute abundances. This species, like the checkerspot butterfly studied by Ehrlich and his colleagues, is population rich. It was these observations that we (Iles and Sinclair) account for. In the Gulf of Maine area, observations of spawning locations and subsequent larval distributions suggest strong links to the physical geography. Several of the populations (Georges Bank, Nantucket Shoals, Grand Manan, and southwestern Nova Scotian) spawn in the late summer to autumn close to the predicted positions of tidally induced fronts (i.e., at transition values of the stratification parameter, H/u^3, where H is the depth and u is the mean tidal current velocity). Simpson and Hunter (1974) have demonstrated that in tidally energetic seas, such as the North Sea and the Gulf of Maine, during the months from spring to autumn, different values of the stratification parameter accurately describe areas that are vertically well-mixed and thermally stratified. Also, the geographic locations of transition from well mixed to thermally stratified water columns, the location of persistent temperature fronts, are associated with a critical value in the parameter. This simple approach (i.e., the analysis of the distribution of H/u^3) has aided considerably in the understanding of persistent spatial structure in the water properties in tidally energetic seas.

Atlantic herring life cycles in such seas are associated with aspects of this structure. The larval surveys in the Gulf of Maine area indicate that the larval distributions persist for several months, mostly within the vertically well mixed areas. During the long larval phase in this area (autumn to the following spring), the frontal structure breaks down completely, indicating that the fronts themselves are not critical to the retention process. The spawning locations, larval distributional areas, and the distribution of the stratification parameter for the Gulf of Maine area are illustrated in Figures 3.1 and 3.2.

Observations from other seas whose tidal processes are important in defining circulation and mixing suggest that some other herring populations are associated with similar physical oceanographic features (Figures 3.3, 3.4, and 3.5). However, herring populations and spawning locations are found in geographic areas that are not characterized by

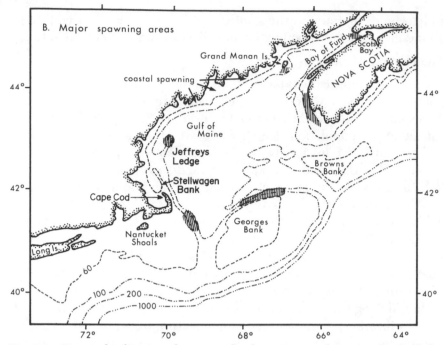

Fig. 3.1 Geographic location of spawning for the major populations in the Gulf of Maine area. (Reproduced from Sinclair and Iles 1985a)

strong tidal circulation. Such spawning populations have been observed in the inland sea of Cape Breton, Nova Scotia (the Bras d'Or Lakes population or populations), the estuaries along the coast of Maine, the coastal embayments along the west coast of Ireland, and the coastal currents along Norway and around Iceland. In many of these examples the spawning locations and subsequent larval distributions are associated with well-defined and geographically predictable or stable oceanographic (or simply geographic) systems such as estuarine circulations, coastal currents, and semi-enclosed embayments. The common denominator for the diverse spawning locations of the separate populations is the geographic predictability of an oceanographic or geographic system that will permit persistence of the larval distributions for a few months after hatching.

The combined observations of herring distributions and physical oceanographic features generated the following hypothesis by Iles and Sinclair (1982):

> The number of herring stocks and the geographic location of their respective spawning sites are determined by the number,

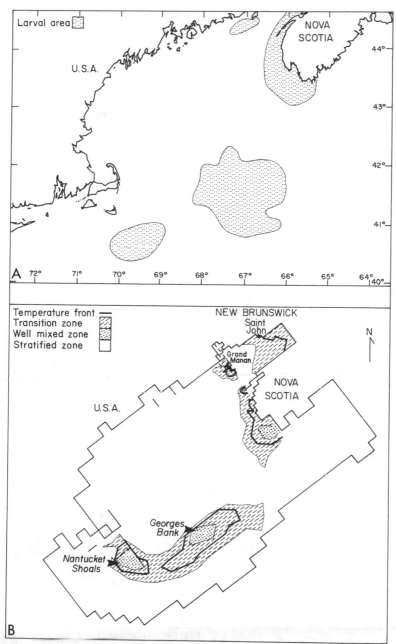

Fig. 3.2 In the Gulf of Maine, relationship of (A) larval distributional areas for several populations of Atlantic herring to (B) location of stratified and well-mixed waters during summer and autumn, predicted using the stratification parameter. (Reproduced from Iles and Sinclair 1982)

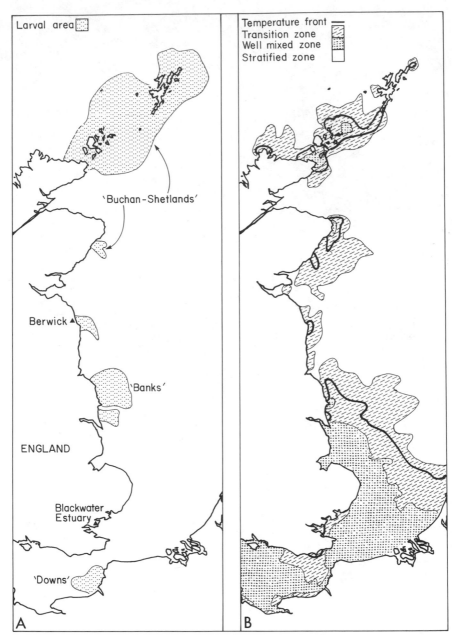

Fig. 3.3 In the North Sea, relationship of (A) early larval distributional areas for several populations of Atlantic herring to (B) location of stratified and well-mixed waters during summer and autumn, predicted using the stratification parameter. (Reproduced from Iles and Sinclair 1982)

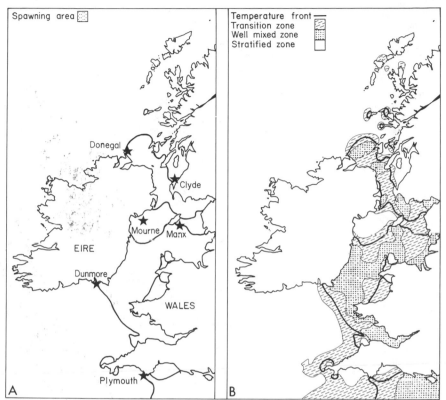

Fig. 3.4 In the Irish and Celtic seas, relationship of (A) spawning locations of several populations of Atlantic herring, indicated by stars, to (B) predicted location of stratified and well-mixed waters from spring to autumn. (Reproduced from Iles and Sinclair 1982)

location, and extent of geographically stable larval retention areas.

The very existence of a population, we argue, depends on the ability of the larvae to remain aggregated during the first few months of life. In locations where the physical oceanography is such that the larvae from a spawning group with the use of specific behavior (such as appropriately timed vertical migration) can maintain an aggregated distribution, a herring population can be sustained. For herring, which attach their eggs to the bottom, it is crucial that the superadjacent water column feature within which the larvae will develop be geographically fixed, rather than spatially ephemeral, on an interannual time scale.

The herring hypothesis stresses the role of behavior at different

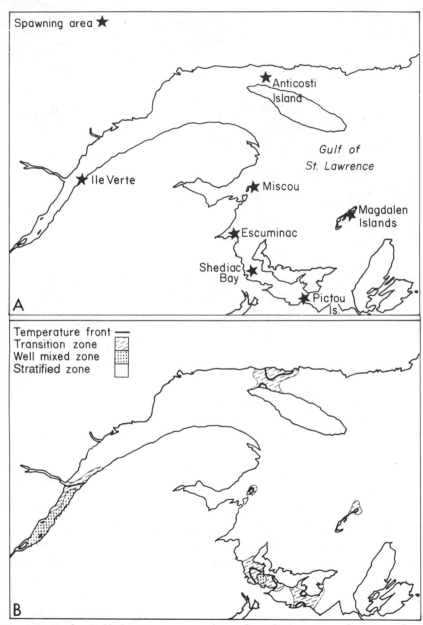

Fig. 3.5 In the Gulf of St. Lawrence, relationship of (A) spawning locations of several populations of Atlantic herring, indicated by stars, to (B) predicted location of stratified and well-mixed waters during summer and autumn. (Reproduced from Iles and Sinclair 1982)

stages of the life cycle in relation to physical oceanographic constraints. Homing must be to precisely defined locations associated with geographically stable water column circulation and mixing characteristics. The hypothesis (contrary to the treatment of the egg and larval phases by Fulton, Damas, and Harden Jones summarized in Chapter 2) emphasizes active retention based on behavior in relation to the complex vertical and horizontal circulation and mixing characteristics of the larval distribution area, rather than passive drift with the surface-layer residual circulation. It is not, however, a larval retention hypothesis but a hypothesis dealing with the existence of herring populations, in particular their spawning locations. It attempts to account for the observed richness and the particular pattern of populations of Atlantic herring. On ecological time scales, relative persistence in the spatial patterns has been observed. Yet many of the spawning locations within the Gulf of Maine area were above sea level several thousand years ago. The hypothesis addresses the ecological time scale but not the ability to establish populations in new areas on the geological time scale.

The additional major question on Atlantic herring that Iles and Sinclair (1982) address is, what controls the characteristic levels of absolute abundance for the different populations (Table 3.1)? Abundance is observed to be at least a partial function of the physical oceanographic features of the superadjacent water column in the spawning area. Iles and Sinclair argue (p. 630) that population abundance is a function of the size of the physical system underlying larval retention (Figure 3.6): "Small stocks are associated with small hydrographic features (for example, the Ile Verte area of the Gulf of St. Lawrence . . .), and large ones with large features (for example, the Georges Bank well-mixed zone . . .)."

If population abundance is defined in relation to processes at the spawning site, and to the egg and larval phases, the carrying capacity for juvenile and adult herring may not be reached. Juvenile annual growth rates in the Gulf of Maine area have been shown to vary inversely as a function of abundance (Sinclair et al. 1981). There is, however, little evidence that food is generally limiting for adult herring. Hjort (1934) made this generalization at the Huxley Lecture at the University of London:

> It is interesting that the same environment should be able to support such varying numbers. A single year class of the Norwegian herring has been found to comprise thirty times as

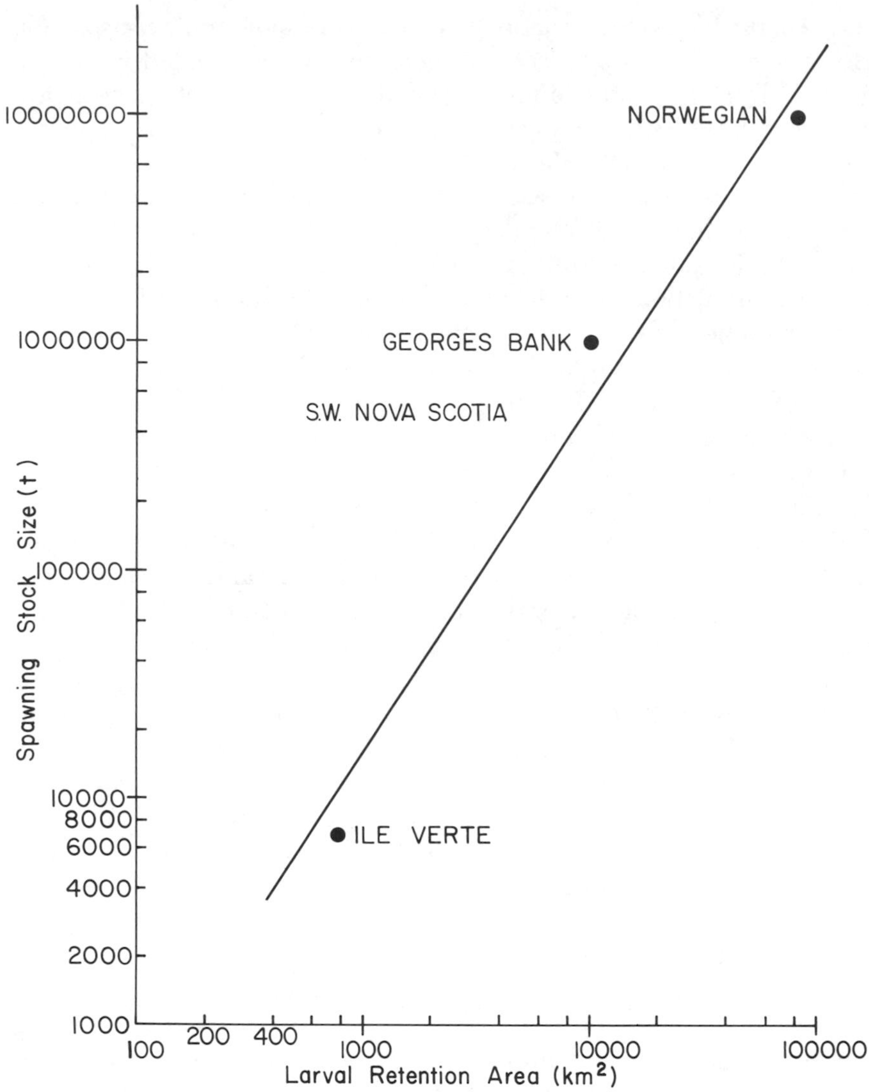

Fig. 3.6 Estimated biomass of several populations of Atlantic herring as a function of the size of the physical oceanographic feature inferred to be associated with the larval retention area. (Reproduced from Iles and Sinclair 1982)

many individuals as the others. . . . In many years there would seem to be a vast surplus of the means of subsistence in relation to the stock of fishes.

Annual growth rates for adult herring of different ages off south-western Nova Scotia have been observed to increase rather than decrease as a function of abundance over a broad range (Sinclair et al. 1982). At very high abundance levels, however, some evidence of reduced growth rate was observed. Haist and Stocker (1985) have drawn the same conclusion for Pacific herring. In addition, the interannual variability in herring fecundity in relation to adult weight is low in spite of high year-class variability (Ware 1980). If absolute abundance is ultimately a function of physical oceanographic processes at the spawning locations (as inferred in Figure 3.6), there is no requirement for limitation by food-chain processes at other phases of the life history in order to have relative stability in numbers over time. Based on this argument, it is not, however, to be excluded as a controlling factor.

Cushing's Critique of the Herring Hypothesis

The Atlantic herring hypothesis of Iles and Sinclair has been criticized by Cushing (1986). He addresses two separate issues: (1) the inapplicability of Iles and Sinclair's hypothesis to the eastern Atlantic; and (2) the reasons for the differences in numbers of populations of gadoids in the eastern and western Atlantic. Only the first question is addressed in this section. (The second question, however, is addressed in the essay as a whole.) Cushing concludes that "[from] two lines of evidence, the hatching of herring eggs often outside the period of stratification and the long larval drift, the Iles and Sinclair hypothesis cannot be extended to the waters around the British Isles." Since the herring hypothesis forms the basis for much of the conceptual developments in this study, Cushing's evaluation is discussed in some detail.

In our view, a major problem with Cushing's critique, and the resulting conclusion just quoted, is a misunderstanding of the hypothesis itself. The hypothesis does not infer that eggs from all, or even most, spawning populations should be hatched during periods of thermal stratification; nor is displacement of larval distributions from the location of spawning necessarily inconsistent with the hypothesis. It is important to emphasize that it is not a larval retention hypothesis but rather a hypothesis to account for the number of herring populations, their specific location of spawning, and their mean absolute abundance. Perhaps we were not sufficiently clear in defining a "geographically stable larval retention area." We mean by this term that there has to be a water-column physical oceanographic (or geographic) system that exists in the same geographic location from year to year. Herring populations, we hypothesize, lay their eggs (which are then attached to the bottom) in relation to such fixed water-column circulation systems.

In this sense the term "larval retention area" refers to a physical oceanographic phenomenon. Circulations associated with tidal processes, because of the regularity in the spatial distribution of the dissipation of tidal energy, are in this sense "geographically stable." We infer that such circulation systems (which in a three-dimensional sense are not well understood by physical oceanographers) are important in the definition of herring populations. It is only in tidally energetic areas, however, that the circulations associated with tides themselves are a dominant component of the overall circulation. In areas not characterized by high tidal energy, other physical processes (e.g., density gradients, wind, and geographical constraints) may or may not generate geographically stable circulations. In sum, we hypothesize that each herring population spawns at the edge of a water-column physical oceanographic system that is, geographically speaking, persistent from year to year. This system, coupled with the behavior of the larvae themselves, enhances coherence in the larval distribution during the first few months of the early life history.

In this sense—that is, from a consideraton of life-cycle closure in geographic terms—the hypothesis emphasizes retention rather than drift at the larval phase. Scale is important. Individual larvae, of course, do drift, and the center of distribution of the larvae from a given spawning population may be displaced in the downstream direction by the residual current. However, it is the minimization of dispersion and transport that is considered critical rather than the drift. The larvae (in anthropomorphic terms) are not trying to drift with a residual current to a particular nursery area, as hypothesized by Fulton (1889), Damas (1909), Schmidt (1909), and Harden Jones (1968), but are doing the exact opposite: trying to maintain an aggregated distribution in spite of the diffusive state of the physical environment. We are indeed saying something different from Harden Jones's migration triangle, but we are not claiming that larvae have anchor lines to the bottom. Older larvae, or recently metamorphosed juveniles, are inferred to migrate actively to their juvenile nursery grounds (sometimes against the residual current) rather than to drift passively with the currents.

Iles and Sinclair's herring hypothesis addresses the geographical or spatial basis of herring populations and the control of their mean abundance. It generates considerable explanatory power. No other hypothesis accounts for *both* the geographic basis of populations (why do herring spawn where they do?) and their characteristic absolute abundances. Blaxter (1985), perhaps unaware of the hypothesis, has subsequently made this point (i.e., that there is no hypothesis to account for differences in abundance between populations).

In addition, the hypothesis has generated an interpretation of

- the timing of spawning of different herring populations (Sinclair and Tremblay 1984),

- the life-cycle distributions of herring in the Gulf of Maine area (Sinclair and Iles 1985), and

- the greater susceptibility of herring management units to recruitment overfishing (Sinclair et al. 1985a).

The first two issues are addressed ahead. This explanatory power concerning a wide range of population phenomena needs to be kept in mind when evaluating the claim that the hypothesis is inapplicable to the eastern Atlantic. In the papers just mentioned, consistent observations concerning the Atlantic herring were noted throughout its distributional range. Given the fundamental nature of our herring hypothesis, if it is found to be inapplicable in the eastern Atlantic, as claimed by Cushing, it is no doubt inapplicable throughout the distributional range of the species.

We shall now deal directly with the evidence put forward against the hypothesis. It is argued by Cushing that hatching for many autumn-spawning populations around the British Isles occurs after the tidally induced fronts have broken down. We do not contest these observations. However, Iles and Sinclair (1982) never claimed that the temperature fronts themselves are critical or essential to defining larval retention areas. For those herring populations that spawn in tidally energetic areas we argued that the local features of the tidal circulation are (with behavior) the critical requirements for the existence of a population and the definition of the precise location of spawning (see pp. 628–629 of Iles and Sinclair 1982). It is not necessarily the temperature discontinuities that are important in defining the location of spawning, but rather a circulation system that enhances discreteness of the larval distribution. Temperature fronts are given a possible role in some cases (by default because of a lack of other circulation features), but it is recognized that during the winter they do not exist. Iles and Sinclair (1982, fn. 26) indicate that due to the weak tides in the Gulf of St. Lawrence the stratification parameter may reflect, not temperature discontinuities, but relative degrees of tidal mixing. They also note

(1982, fn. 27) that tidally induced circulations persist during the winter months when the temperature fronts have broken down. Some features of the tidal circulation occur throughout the year, although they are enhanced by the horizontal gradients in temperature (and thus density) during the summer months. We did point out that circulation characteristics contiguous to the spawning locations in the central (Bank) and southern (Downs) North Sea were not clear. Since that time there has been an increased understanding of the physical oceanography of the North Sea. This issue is discussed ahead.

For populations that do not spawn in tidally energetic seas we argued that other aspects of the physical geography are critical to the retention of larvae during the first few months of life (such as estuarine circulation, persistent gyres, and the bounded nature of coastal embayments themselves). In short, the observations that larvae hatch in areas that at times are not characterized by temperature fronts are not evidence against the herring hypothesis. Perhaps this point was not clearly enough underlined in the original presentation. The several illustrations themselves (Figures 3.2–3.5), linking the location of spawning and of larval distributions to predicted tidally induced features, may have left the impression that the temperature fronts are the features of importance to the definition of spawning location. It is clear in the text, however, that it is the circulation features associated with the H/u^3 distributions that are considered more important.

The second set of observations put forward as evidence against the herring hypothesis are descriptions of larval distributions in the North Sea and their changes through time. The populations considered are the so-called Downs, Bank, and Buchan herring. As a preliminary point we do not agree that these three herring units are in reality the self-sustaining populations in the North Sea. Each unit may well comprise several populations. Bank herring, for example, are probably a population complex which includes separate populations spawning on the Yorkshire coast and on Dogger Bank itself (in past years). From the evidence of persistent spawning in particular geographic areas that are coincident with H/u^3 and other physical features, one can infer (following our hypothesis) that there are considerably more than three populations of herring in the North Sea. In other words, there are more than three larval retention areas. The grouping of the populations into three units may have been convenient for management purposes, but it has perhaps masked the distributional complexity and population richness of this species and hindered progress toward a real understanding of the population structure in the North Sea.

In the remainder of this section the evidence presented in Cushing's critique for larval drift in the North Sea is evaluated. Particular attention will be focused on the southern North Sea because the evidence against our hypothesis is perhaps the strongest in this geographic area.

When does a set of observations support larval retention rather than larval drift? This is not an easy question, as we shall see. As indicated above, we are not suggesting that larvae do not move from their location of hatching. Rather, we infer that successful hatching (in the sense that the eggs contribute to a self-sustaining population) occurs only in geographic locations that are contiguous with a circula tion system that permits, with the behavior of the larvae, aggregation of the distribution for the first few months of the larval phase. We suggest that the Norwegian coastal current itself is a larval retention area (albeit the largest example). Subsequent to hatching, early-stage larvae have been observed to drift very close to the bottom for several days before they begin to migrate vertically within their larval retention area (which is particularly well studied for the Ile Verte population in the upper St. Lawrence estuary). Thus, drift itself (either within the Norwegian Current or in the bottom water of the St. Lawrence) is not inconsistent with the concept of larval retention.

This does not, however, lead to the conclusion that the latter concept is without meaning or rigor; nor does it mean that the herring hypothesis is the same as the migration triangle of Harden Jones. Both emphasize opposite processes at the planktonic stage of the life cycle. The reason for the difference in emphasis is partially a result of a shift, between the 1960s and the 1980s, in the perception of the continental shelf circulation. In the earlier period large-scale residual surface-layer circulations were considered meaningful, whereas in some continental shelf areas it may be that the variability (particularly if a three-dimensional image is considered) is of greater significance to eggs and larvae than the mean.

Cushing (1986, Fig. 1) reproduces Bückmann's summary description of the displacement of the herring larval distribution in the southern North Sea (from west to east) in the direction of what was then described as the surface-layer residual current. The average currents in this area, as understood at that time by fisheries biologists, are illus-

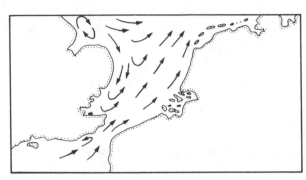

Fig. 3.7 Surface-layer residual circulation of the southern North Sea as perceived in the 1950s and 1960s. (Redrawn from Harden Jones 1968)

trated in Figure 3.7 (taken from Harden Jones 1968). Bückmann's study is one of the best in the literature for all marine fish species, not just herring, in support of the concept of larval drift and Harden Jones's migration triangle.

The physical oceanography of this area is becoming better understood. Ronday (1975) and Nihoul and Ronday (1974) describe the effects of the tidal stress on the residual circulation in the southern North Sea. Their model suggests a southwesterly current near the Belgian coast associated with a gyre. Pingree and Maddock (1985), in a more sophisticated analysis, also conclude that there is a tidally induced gyral circulation off the coast of Belgium and the Netherlands, shown in Figure 3.8. The illustration is complex and perhaps requires some interpretation. The terms "Eulerian circulation," "Lagrangian circulation," and "Stokes drift" are related by the authors (p. 970) as follows: "To relate the mean velocity *following* the flow (Lagrangian description of the flow pattern) to the mean velocity at a *fixed point* (Eulerian description of the field of flow) it is necessary to add the so-called 'Stokes drift' to the Eulerian mean velocity" (emphasis added). The upper half of Figure 3.8 shows the Stokes transports and Eulerian currents.

In their text Pingree and Maddock (p. 978) point out that the Southern Bight of the North Sea is filled through the Strait of Dover and the western side of its northern boundary (arrows 4 and 6 in Figure 3.8). Not much of this Stokes transport leaves the Southern Bight through the central and eastern sides of the northern boundary. They explain that, from consideration of continuity of volume, the net Stokes flux is converted into Eulerian circulation, which is directed to the north-northeast through the central region of the northern boundary. In other words, most of the transport, irrespective of the season, is to the west of the larval distributional area of the herring spawning at the

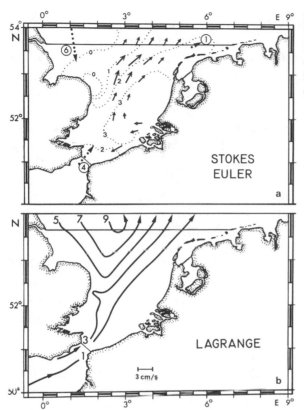

Fig. 3.8 Residual tidal transports in the southern North Sea (see text for discussion). (Reproduced from Pingree and Maddock 1985)

mouth of the Southern Bight. In fact, the Eulerian currents along the coasts of the Netherlands and Belgium are to the west (i.e., toward the Strait of Dover). The Lagrangian transports in the lower part of Figure 3.8 indicate weak transport to the northeast in the larval distributional area. The circulation described by Pingree and Maddock is very different from that available to Harden Jones (compare the Southern Bight circulations in Figures 3.7 and 3.8).

Unlike Cushing, we find, given the increased understanding of the physics of the area, considerable support for the herring hypothesis from the empirical observations in the southern North Sea. The tidally induced residual Eulerian circulation, we suggest, is *part* of a circulation system that enhances coherence in the larval distribution for the population spawning to the east of the Strait of Dover. We emphasize that this circulation is only part of a larval retention system; the model does not include depth differences in circulation. Since herring larvae after the first several days begin daily vertical migration, the depth profile in circulation and mixing is no doubt important in influencing the distribution of the larval population. Herring spawning to the west of the Strait of Dover off the coast of France could be a separate population with its own larval retention area.

Several data sets on spawning locations and larval distributions in this geographic area (summarized in Bridger 1961; and even Bückmann 1950) support the herring hypothesis. First, spawning occurs at the same time of year in a specific geographic location year after year. The recent resurgence of the herring population(s) in the southern North Sea has taken place in the identical geographic locations observed prior to the collapse in the early 1960s. Second, the gyral circulation in the area contiguous to spawning is geographically stable since it is tidally generated. Third, the larval distributions, as we interpret them, are retained in this physical system.

Of these studies, Bridger's (1961) provides the most complete coverage of the geographic area during the first few months of the larval phase (Figures 3.9 and 3.10). The station density is relatively high, but the geographic area covered varies between years. Two observations are of interest:

1. The contours during the months of January and February (particularly clear during 1957 and 1958 samplings) indicate that the larval distribution is partially coincident with the gyral circulation of Pingree and Maddock (1985) shown in Figure 3.8. (Unfortunately, the sampling grid rarely extends close to the coasts of Belgium and the Netherlands in the earlier years of the program.)

2. There is limited displacement of the distribution toward the northeast (compare January to February 1957 and 1958 in Figures 3.9 and 3.10). During three months a conservative estimate of the displacement due to the residual circulation in the southern North Sea is 500 km. The observed displacement of the center of the larval distribution from Bridger's summary work is less than half that predicted on the basis of larval drift.

Bückmann's (1950) observations are more difficult to evaluate because of the way he presents the data, which were prepared to support the contemporary larval drift interpretation of the larval phase. A further complication to the interpretation of Bückmann's observation is the existence of spawning populations off the German coast itself. Bückmann (p. 53) himself indicates the existence of a gyre off the coast of the Netherlands:

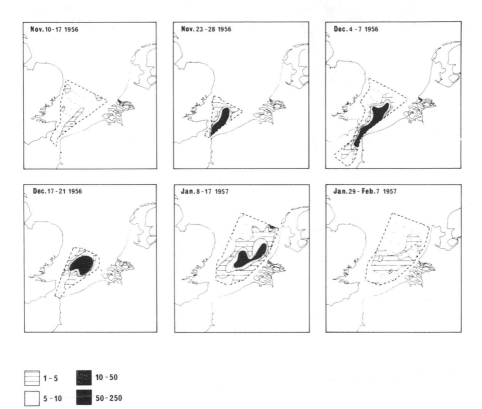

1 - 5   10 - 50
5 - 10   50 - 250

Fig. 3.9 Herring larval distributions in the southern North Sea during the winter of 1956/57. (Redrawn from Bridger 1961)

> Early in January the bulk of the larvae that hatched in November are found off the coast of the Dutch province of Zeeland. Some observations indicate that such larvae remain behind there for some time, caught in a little eddy off the mouth of the Rhine. But as a whole the transport or movement proceeds in a northeasterly direction.[3]

Given what was known in 1950 about circulation and the extant hypothesis of larval drift, it is not surprising that Bückmann, in his overview, emphasizes the displacement of the distribution to the northeast rather than the slowing down of that displacement by a gyre. He entitled his figure "Spreading of herring larvae out of the Flemish Bight." One could entitle the same figure "Retention of herring larvae within the Flemish Bight with some evidence of leakage." From Bridger's more detailed coverage and the improved understanding of the circulation, a larval retention interpretation provides an alternative conceptualiza-

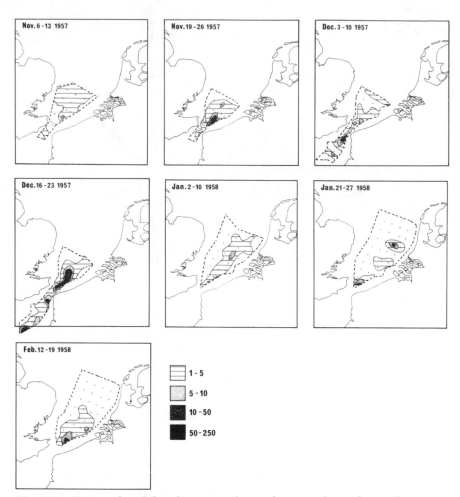

Fig. 3.10 Herring larval distributions in the southern North Sea during the winter of 1957/58. (Redrawn from Bridger 1961)

tion of the observations that is consistent with the observations in the northwestern Atlantic.

Talbot (1974), from an analysis of the accumulated plaice egg and larval distributions in the southern North Sea, estimates that the distributions do not change over time according to the predictions of the physics of mixing in the area. The distributions do not disperse to the degree predicted for a passive contaminant. An alternate interpretation is that the animals are actively doing something to minimize the dispersal of their distribution. It is this active process of retarding dispersal and transport that we infer is occurring in the gyral circulation in the southeastern North Sea. In this particular case, if looked at in isolation from herring populations, a larval drift interpretation is not inconsistent with the empirical observations. However, we believe that it is the wrong interpretation. The issue becomes one of interpreting the evidence in relation to two different conceptual frameworks dealing with complex life histories in the marine environment.

The second geographic area dealt with by Cushing (1986) is the so-called Bank herring spawning locations and subsequent larval distributions in the central North Sea. A careful reading of Bückmann (1950) does not support the conclusion by Cushing that the Bank larvae "are drifted steadily across the North Sea." Bückmann's summary of the available data (1950, p. 54), again interpreted within the context of the larval drift hypothesis and the available overview of the residual circulation of the central North Sea, is both cautious and complicated. He concludes tentatively that Dogger Bank larvae (from whatever source) may seek the coast of England for their metamorphosis rather than drift across the North Sea. This is a far cry from Cushing's representation of Bückmann's analysis. There would seem to be room for alternative interpretations of the empirical results that emphasize retention using the complex circulation rather than drift with the surface-layer residual.

Other central North Sea observations in the literature can be argued to support the concept of larval retention. Zijlstra (1970) describes herring larval distributions in the central North Sea at a time for several years when the Dogger spawning population was still healthy. Two points are of interest. Prior to the collapse of the spawning population on southwestern Dogger Bank there is some evidence of discreteness between the larval distributions resulting from this population and one spawning off the coast of Yorkshire. During the year when the station density was highest (1961) the discontinuity (to the degree that it actually existed) is the most clearly defined. In that year the discontinuity was clearly observed for the larger larvae (11–15 mm), as well as for the recently hatched larvae.

Second, from the similarities in the distributions of the two size categories for a given year it can be deduced that there is a persistence (or retention) of larvae within a particular area over the period of growth. Many past interpretations have attributed the presence of larger larvae in the proximity of a spawning location to drift of these larvae from a distant spawning location. An alternate interpretation of the evidence, and the one we favor, is that the larvae have maintained a fairly fixed position in spite of the surface-layer residual circulation by vertically migrating in relation to the short-term variability in the complex three-dimensional circulation.

Other studies in the central North Sea can similarly be interpreted to support the concept of larval retention rather than larval drift. Iles and Sinclair (1985) have provided some evidence for a larval retention area off the coast of Yorkshire.

Cushing cites the study by Clark (1933) as support for passive larval drift from the several populations spawning off the coast of Scotland in the northern North Sea (the so-called Buchan stock). This study, like Bückmann's, is difficult to evaluate. The sampling density and temporal coverage during the early life-history stages of the various herring populations are inadequate for the purposes of Cushing's paper. It was very much a broad-brush approach. From the distribution of larvae with scales, Clark concludes that larvae from autumn-spawning populations metamorphose in the western North Sea rather than drift across the North Sea to continental Europe. Cushing disagrees with this conclusion and places his emphasis on the dispersal and drift of the larvae away from the various spawning locations.

Iles and Sinclair (1982) suggest that the tidal circulation in the vicinity of the various autumn-spawning areas to the north and east of Scotland are critical to larval retention. In our view the observations of Clark (1933) are not sufficiently detailed to test our hypothesis. The data may be available from the diverse larval surveys to evaluate more critically the hypothesis for the northern North Sea. It would seem premature to reject it on the basis of Clark's analysis.

Earlier studies in European waters on the growth of herring larvae have either assumed or demonstrated larval retention. Meyer (1878) observed retention of herring larvae in the Schlei fjord. Because they were retained, he was able to estimate their growth throughout the larval phase. Similar studies on retained larvae have been carried out by Masterman (1895) off St. Andrews and Bowers (1952) off the Isle of Man. In each case the authors were confident that they were sampling a larval population through time at a *fixed* geographic location. In sum, in several locations in the eastern Atlantic, growth studies have been carried out on the unstated assumption that larval populations persist for up to several months in a relatively fixed geographic location.

On the basis of the studies cited by Cushing (1986) we feel that it is premature to reject the herring hypothesis of Iles and Sinclair (1982). It is our view that a major problem in fisheries biology, one that contributes to the difficulty of considering the herring hypothesis seriously for the North Sea, is the image of surface-layer residual circulations as real phenomena. In many continental shelf areas the estimated residual circulation is not a meaningful concept. The short-term variability, and the changes in speed and direction with depth, can be much larger than the estimated residuals. We feel that the larvae may be much more influenced by, and adapted to, the variances than by the mean. The physical oceanography that they are experiencing is not that represented in the oceanographic literature describing average currents, not even in the more sophisticated analyses such as those by Pingree and Maddock (1985).

From hindsight it may be that surface drift card and bottle studies in the North Sea, initiated by Fulton in Scotland at the end of the nineteenth century to test his larval drift hypothesis (see Figure 2.3) and carried on actively under the direction of Garstang and later Graham at Lowestoft, have hindered rather than helped our understanding of the circulation patterns and larval fish ecology. Fisheries biologists have tended to believe that the path of a surface drift bottle tells the direction that fish larvae will drift. We strongly caution against this basic assumption. Surface drift patterns may poorly reflect the actual net horizontal advection of a larva as it migrates daily up and down through parts of the water column. Dooley and McKay (1979), using circulation information from current meters rather than drift bottles, have carried on this tradition of inferring drift from average current velocities at a particular depth. They estimate that herring larvae from the west coast of Scotland could drift into the North Sea in 40 days. We again caution against this kind of exercise. It ignores the empirically observed behavior of the larvae and oversimplifies the relevant physical oceanography.

Life-History Distributions in the Gulf of Maine

The distributional aspects of the life cycle of Atlantic herring in the Gulf of Maine have been described by Sinclair and Iles (1985). The adult annual migration can be pieced together by linking the static descriptions of the summer-feeding, spawning, and overwintering distributions (Figure 3.11). Essentially all the adults from the diverse populations in the Gulf of Maine area are interpreted to feed during the

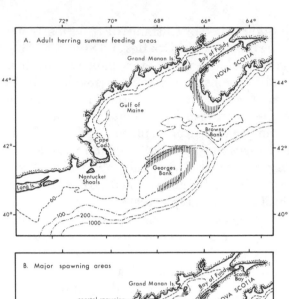

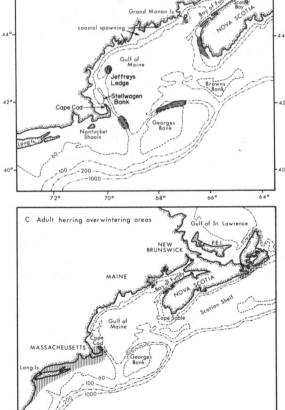

Fig. 3.11 Distributions of Atlantic herring in the Gulf of Maine and Scotian Shelf area during the summer feeding, spawning, and overwintering phases of the adult annual migration cycle. (Reproduced from Sinclair and Iles 1985a)

summer months in only two or three locations (in the vicinity of the highly productive tidally induced fronts). From the tagging results it is found that herring from different populations can share the same summer-feeding environment. Depending on the particular timing of spawning for a population, herring depart from the feeding area and migrate to their respective spawning sites. This does not appear to occur simultaneously. For example, the Scots Bay population in the upper Bay of Fundy begins spawning in July and thus leaves the feeding area earlier than the herring that spawn in the late summer and autumn. Presumably the Scots Bay herring continue to feed after spawning. The several spawning locations are illustrated in Figure 3.1. After feeding and spawning, the adult herring of essentially all the populations in the Gulf of Maine area migrate to two or three overwintering locations (Figure 3.11). The timing of the migrations to and from spawning, summer-feeding, and overwintering areas is remarkably regular from year to year.

The early life-history distributional areas, to the degree that they can be pieced together, are summarized in Figure 3.12. As already indicated, for several months after hatching the larval distributions from the several spawning populations are discrete. The larval phase is quite long for these populations that spawn from late summer to autumn, with metamorphosis not occurring until the following spring. The larval phase covers a part of the year during which food is at a low level (midwinter). The period from the postlarval stage to Age 1 (i.e., the first summer) is not well understood. However, the juvenile nursery areas, again for essentially all herring in the Gulf of Maine area, are in the coastal zone along the shores of Maine, New Brunswick, and Nova Scotia (Figure 3.12). The tagging results indicate that there is also mixing between populations during the juvenile phase of the life history.

What do these simple spatial patterns imply for population regulation? First, it is clear that, except during spawning itself and the first few months of the larval phase, populations mix extensively with each other. The spawning populations may be of very different sizes. The Scots Bay spawning population is much smaller than the population spawning off southwestern Nova Scotia (probably orders of magnitude smaller). Individual herring can belong to populations which are at either the high or the low end of their range of abundance, yet they share a common environment for much of the year. Absolute abun-

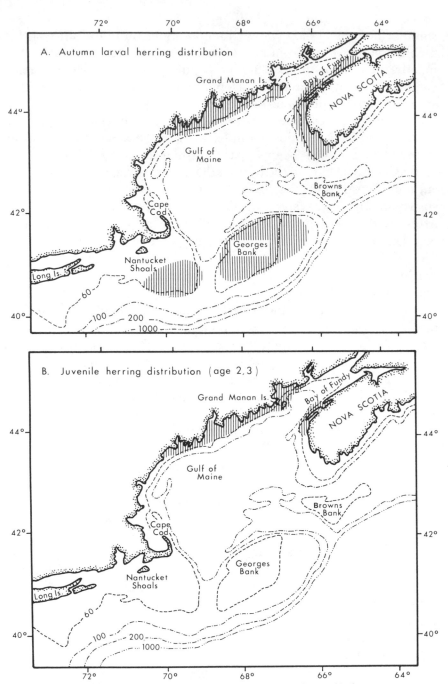

Fig. 3.12 Distribution of larval and juvenile herring in the Gulf of Maine area. (Reproduced from Sinclair and Iles 1985a)

dance of the individual populations, Iles and Sinclair (1982) argue, is defined at the relatively short discrete phase of the life history (from spawning through the first few months of the larval phase). These distributional observations, as well as the herring hypothesis, suggest that intraspecific competition for limited resources during most of the life history cannot be important to population regulation.

A second conclusion is also stressed. In Chapter 2 it was pointed out that the population hypothesis of Harden Jones, involving spawning in relation to egg and larval drift in the surface-layer residual circulation (the migration triangle), does not provide explanatory power. The complex migrations of herring in the Gulf of Maine area, as well as the spawning population patterns, cannot be interpreted in relation to the larval drift/migration triangle hypothesis. The Georges Bank progeny, which as juveniles aggregate in the coastal zone of the inner Gulf of Maine, for example, would in some cases have to migrate against the residual current. The known spawning locations, juvenile nursery areas, and well-described residual circulation in the Gulf of Maine area (Figure 3.13) cannot be pieced together consistently to support the hypothesis based on larval drift and residual currents.

Time of Spawning

As noted above, the time of spawning for particular populations within the Gulf of Maine is well defined and is population specific. Most, but not all, populations in this particular area spawn in the late summer to autumn. If the overall distributional range for the species is considered, the pattern is remarkable. A population is spawning somewhere in the northern Atlantic every day of the year, yet the mean time of spawning for each population is precise and the duration is brief (Figure 3.14 and Table 3.2).

An explanation of this pattern for the species has been the focus of theoretical developments in fisheries biology. Sinclair and Tremblay (1984) have summarized the coupling of the hypothesis of the time of spawning of herring to a hypothesis of recruitment variability (i.e., Cushing's match/mismatch theory, discussed briefly in Chapter 2). The match/mismatch theory states that the time of spawning is adapted to the particular time of spring and autumn phytoplankton blooms in the geographic area of inferred larval drift. Iles and Sinclair (1982) note, however, that phytoplankton seasonal cycles in the larval retention areas of three autumn-spawning populations are not characterized by autumn blooms. Phytoplankton seasonal cycles in both tidally induced

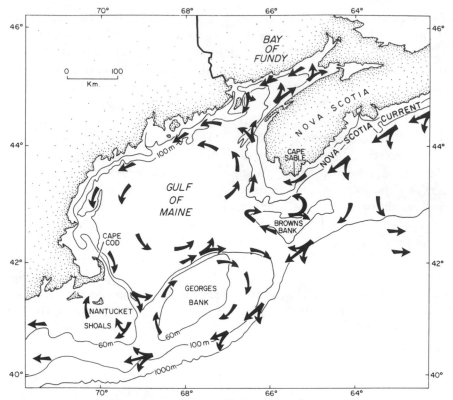

Fig. 3.13 Residual circulation of the Gulf of Maine area. (Drawn by R. Trites, Maine Ecology Laboratory, Bedford Institute of Oceanography, Dartmouth, Nova Scotia, Canada)

frontal areas and vertically well mixed areas are not characterized by a midsummer reduction and subsequent autumn bloom. Yet, herring populations characteristically spawn from late summer to autumn in tidally energetic seas such as the Gulf of Maine and the North Sea.

On this basis alone Iles and Sinclair (1982) and Sinclair and Tremblay (1984) have argued that the conceptual base of both Cushing's hypothesis of time of spawning for herring and his match/mismatch theory is flawed. Further, the hypothesis for time of spawning does not consider the empirical observations, well documented since the 1880s, of marked differences in the duration of the larval phase between different herring populations. Autumn-spawning populations have a larval phase exceeding six months, while this phase for spring-spawning populations can be as short as two or three months. A

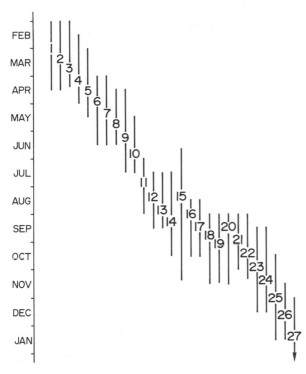

Fig. 3.14 Spawning times of 27 populations of Atlantic herring. Populations are identified in Table 3.2. (Reproduced from Sinclair and Tremblay 1984)

hypothesis on the time of spawning should account for such inter-populational differences in life-cycle patterns.

Sinclair and Tremblay (1984) propose an alternate hypothesis on the time of spawning of Atlantic herring populations, in which timing is a function of the length of time necessary for larval development. Two constraints to herring populations are considered important:

1. Larval phase occurs within geographically well defined larval retention areas which differ between populations in their plankton production characteristics.

2. Metamorphosis from the larval to the juvenile stage can occur only within a restricted time of the year (April to October).

The first constraint is based on the herring hypothesis described above. The second is based on empirical observations.

Three elements support the hypothesis:

1. Timing of metamorphosis is seasonally restricted (Figures 3.15 and 3.16).

Table 3.2 Peak spawning periods for herring populations shown in Fig. 3.14, estimated from peak hatching data (with estimated incubation time subtracted) or from peak catches from a spawning fishery. If observations extend over more than one year, years are shown in parentheses. References are listed in Sinclair and Tremblay (1984).

Population	Peak period
1. Clyde Sea	February 20–28
2. Norwegian	February 18–March 18 (27 years)
3. Minch	March*
4. Blackwater estuary	April*
5. Schlei Fjord–Kiel Bay	April*
6. Magdalen Island	May 9 (27 years)
7. Southwestern Gulf of St. Lawrence	May 14–18 (27 years)
8. Chedabucto Bay	May*
9. Southeastern Gulf of St. Lawrence	May 29–June 6 (26 years)
10. St. Lawrence estuary	June 15–July 7 (5 years)
11. Scots Bay	July 20–August 3 (6 years)
12. Northwestern North Sea	August
13. Southern Gulf of St. Lawrence	July 31–September 13 (27 years)
14. Southwestern Nova Scotia	August 25–September 10 (5 years)
15. Grand Manan (historical)	July–September
16. Minch	September 5–25
17. Banks and Dogger Bank	September*
18. Manx	September
19. Mourne	Late September–October
20. Coastal Gulf of Maine	
Eastern section	September 15–October 17
Western section	October 1–21 (5 years)
21. Jeffrey's Ledge	September 29–October 25 (2 years)
22. Donegal	October
23. Georges Bank	October 5–23 (8 years)
24. Nantucket Shoals	October 12–November 2 (8 years)
25. Dunmore	September–October, December–January
26. Downs	December
27. Plymouth	January

*Time derived by taking midpoint of spawning period.

2. Duration of the larval phase varies between categories of larval distributional areas (Table 3.3).

3. Zooplankton production varies consistently with the duration of the larval phase (Figure 3.17).

The empirical evidence is substantive for the first two points but sketchy for the third. Part of the problem in evaluating this aspect of the

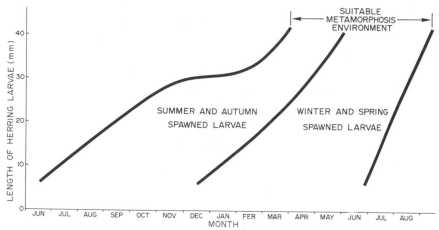

Fig. 3.15 Growth of herring larvae from populations spawning in different months of the year. (Reproduced from Sinclair and Tremblay 1984)

hypothesis (i.e., the differences in feeding environments between larval retention areas) is data comparability. It is difficult to find good comparative studies in zooplankton production, let alone for particular herring larval distributional areas. There is some indirect support from studies of feeding frequency that food availability is lower in the larval retention areas within which the duration of the larval phase is longer (the well-mixed areas in Table 3.3). The observations are noisy (Table 3.4), but a trend is observable. A higher percentage of the larvae are feeding in the shorter-duration larval retention areas.

The summary of the information on duration of larval phase for different herring populations is interesting in light of the accumulated observations of differences in egg size and fecundity between populations. Herring populations differ considerably in fecundity and egg size. In general, the total energy investment in gonads for a herring of a given size is fairly constant between populations, varying by a factor of approximately 2.5, but fecundity and egg size vary by factors of 4 and 5, respectively. Thus, there is an inverse relationship between size of eggs and the number of eggs produced. There are clear patterns in the fecundity-egg size trade-off: spring spawners have larger but fewer eggs, and autumn spawners have smaller but more eggs. A simple interpretation consistent with the time of spawning hypothesis is that, given a poor larval retention area and a concomitant long larval phase, there will be greater accumulative population losses from the distributional area (for the population in question) during the larval phase.

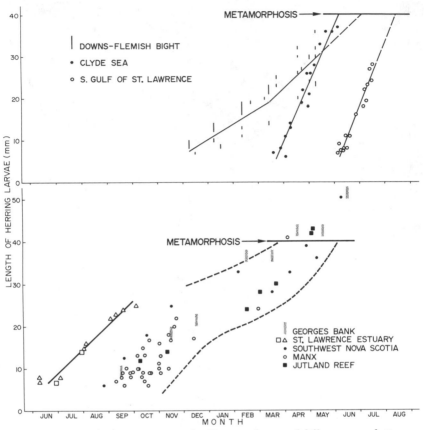

Fig. 3.16 Observations on the growth of herring larvae of different populations. (Reproduced from Sinclair and Tremblay 1984)

Thus, higher fecundity for a given total energy allocation to gonads at the cost of smaller eggs is advantageous. The relationship between duration of the larval phase and fecundity is shown in Figure 3.18.

Tests of the Match/Mismatch Theory

There have been several recent tests of the match/mismatch theory of control of year-class size. According to this theory, population numbers are controlled during the larval phase by the availability of food. As mentioned above, in the light of recent developments in oceaography and fisheries biology, Sinclair and Tremblay (1984) argue that the basic arguments in support of the theory have not stood the test of time.

The new hypothesis just summarized for the spawning time of

Table 3.3 Herring larval phase duration and seasonality of metamorphosis (assumed to begin at 40 mm). Retention area characteristics are from Garrett et al. (1978), Pingree and Griffiths (1978, 1980), and references cited therein. References are listed in Sinclair and Tremblay (1984).

Population	Larval retention area characteristic	Duration of larval phase (months)	Month(s) of metamorphosis
Schlei Fjord–Kiel Bay	Stratified	2.5–3	June–July
Clyde Sea	Stratified	3	June
Norwegian	Stratified	3–4	June–July
Plymouth	Stratified	4–5	April–May
Blackwater estuary	Stratified	4–5	August
Downs	Unknown	6	June
Georges Bank	Well mixed	7	April–May
Manx	Well mixed	7	April–May
Northwestern North Sea (Aberdeen)	Well mixed	7	April–May
Minch	Well mixed	7	April–May
Dogger Bank	Well mixed	7–8	May
Southwestern Nova Scotia	Well mixed	7–8	April–May
Ile Verte, St. Lawrence estuary	Well mixed	10–11	Spring (inferred)

herring decouples the control of this life-history feature from the generation of year-class variability (i.e., the control of population numbers), and accounts for both the seasonally diverse empirical observations on spawning time (a herring population is spawning every day of the year) and the marked between-population differences in the duration of the larval phase (2–11 months). The critique argues that the conceptual development of the theory is invalid, but it does not necessarily demonstrate that the theory is wrong. It could be correct for reasons not stated in its original formulation.

However, five recent field tests do undermine the validity of the match/mismatch theory. Two semi-enclosed regions in the Gulf of St. Lawrence, the lower St. Lawrence estuary and the Bay of Chaleur, were investigated in regard to phytoplankton seasonal cycles, physical oceanographic features, and seasonal spawning events of marine fish. Sinclair (1978) and Lambert (1983) have shown that the two environments have very different seasonal phytoplankton cycles: the Bay of Chaleur has a classical spring and autumn bloom, whereas the lower St.

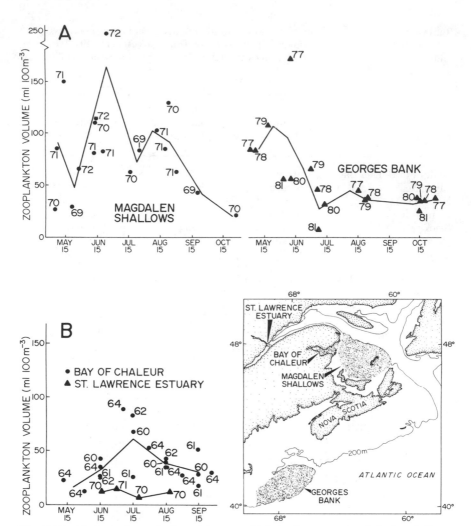

Fig. 3.17 Estimated zooplankton biomass which is characteristic of the larval distributional areas for several herring populations. Bold lines indicate mean conditions over years considered. (Reproduced from Sinclair and Tremblay 1984)

Lawrence estuary is a "late bloomer" (Sinclair et al. 1981) with a broad midsummer peak. However, the spawning times of diverse fish species in the two areas are very similar (de Lafontaine et al. 1984a, 1984b). This comparative study, designed specifically to test the match/ mismatch theory, demonstrates that, at least in these two areas, the time of spawning is not related to the time of seasonal blooms. The

Table 3.4 Feeding incidence of larval herring. Only larvae past the yolk-sac stage that were caught during daylight or during a feeding peak are considered. (References are given in Sinclair and Tremblay 1984.)

Population	Month or season	Larval size (mm)	Percent with food
Schlei Fjord	May	14–20	94–97
	June	19–24	74–100
Norwegian Sea	April	20	84
Clyde Sea	March–April (3 years)	10–20	70[a]
Northwestern North Sea	September	24	65[a]
	September	14–17	50
	October	25	85[a]
	November	34	60[a]
Manx	September	10–15	69
	October	10	70
	November–December	10	64
Gulf of Maine	Autumn	7–30	57
	Winter	11–40	51
	Spring (3 years)	21–50	61
Banks and Dogger Bank	June–November	10–18	50[a]
Georges Bank	October (10 years)	8–12.9	28
	October (10 years)	13–17.9	33
	October–November (3 years)	9–18	21
	December–February (3 years)	15–28	29
	April–May	34[b]	57
St. Lawrence estuary	June	7–8[b]	0–8
	June	10–16[b]	23–53

[a]Peak.
[b]Mean.

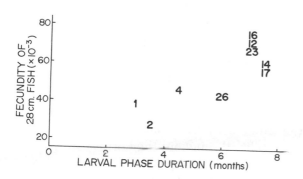

Fig. 3.18 Fecundity of several Atlantic herring populations as a function of the duration of the larval phase. The numbers represent the populations listed in Table 3.2. (Reproduced from Sinclair and Tremblay 1984)

empirical observations, therefore, do not support the theory tested (de Lafontaine 1984a, 1984b).

In the second explicitly stated test, by Methot (1983) on northern anchovy, *Engraulis mordax*, off southern California, the birth dates of juveniles were determined from daily increments in otoliths and compared with the seasonal distribution of spawning determined from ichthyoplankton surveys conducted during the 1978 and 1979 spawning seasons. Methot concludes, in the abstract of his thesis (1981), "The seasonal patterns of survival through the laval stage were not related to larval growth or mortality during the early larval stage but were consistent with changes in offshore drift inferred from monthly upwelling indices." His results, then, support neither the match/mismatch theory of Cushing (1975) nor its particular elaboration for northern anchovy by Lasker (1975, 1978).

The third test is less direct. From broad-scale ichthyoplankton surveys over the Scotian Shelf during most months of the year and covering several years, O'Boyle et al. (1984) observed that spawnings of different species of demersal fish occur throughout the year (Figure 3.19), yet the phytoplankton blooms are seasonal and rather brief. Thus, all species cannot be adapting their time of spawning to the time of phytoplankton blooms, and the match/mismatch theory cannot be generally applicable. It could, however, still be relevant to a subset of the observed species.

The fourth test is also indirect, in that larval mortality itself is not estimated. Koslow (1984) analyzed recruitment time series for 14 populations of northwestern Atlantic fish. Coherence was observed in year-class strengths between populations within a species over very large geographic areas (which was particularly marked for Atlantic cod, *Gadus morhua*). He suggests (p. 1722), "The spatial extent of these patterns, which span the region from west Greenland to Georges Bank, indicates that large-scale physical forcing, rather than local biological interactions, predominantly regulates recruitment to northwest Atlantic fisheries." Koslow, however, indicates that food limitation at the larval phase may regulate recruitment variability, but that there would have to be coherence in the zooplankton abundances over the same extensive geographic area. Further analyses (Koslow et al. 1987) of cod and haddock recruitment variability throughout the northwestern Atlantic in relation to a range of environmental parameters suggest that large-scale physical processes are important, but the mechanisms gen-

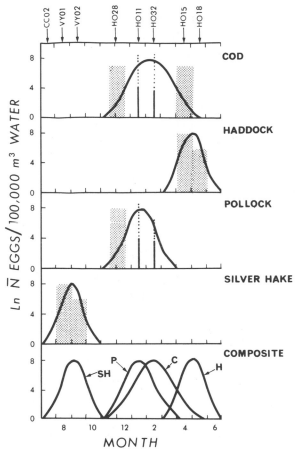

Fig. 3.19 Estimated spawning times of Atlantic cod, haddock, pollock, and silver hake on the Scotian Shelf. The shaded histograms represent egg abundance estimated from data of successfully completed cruises; vertical lines (solid and dotted) represent estimates from incomplete cruises (HO11 and HO32). The bottom panel shows the best estimate of timing of spawning for the four gadoid species that are considered. (Reproduced from O'Boyle et al. 1984)

erating year-class strength coherence between areas remain unclear. We interpret the results as not supporting the match/mismatch theory, in that coherence in the timing of spawning events and plankton blooms would not be expected to occur over thousands of kilometers.

In the fifth test, by Sinclair et al. (1985c) on Pacific mackerel recruitment variability, a survival index for the population (numbers surviving to Age 1/egg production) for the 28-year period from 1938 to 1965 was analyzed in relation to zooplankton abundance. High survival rates were observed during years of low food availability, which is inconsistent with the match/mismatch theory (this study is described in more detail in Chapter 6).

No direct field tests have been made that support the match/mismatch theory, in spite of the considerable efforts that have been directed toward this aim. Support has come, rather, from the experimental work of Laurence (1977) and Lasker (1975, 1978) (whose experimental results were not supported by Methot's) and from interpretative studies in which year-class variability has been statistically related to physical "proxy" variables (Sutcliffe 1972, 1973; Sutcliffe et al. 1977, 1983; Leggett et al. 1984). George and Harris (1985) have interpreted coherence between zooplankton abundance, wind mixing, and temperature observations in Lake Windermere (United Kingdom) as supporting the match/mismatch theory. High zooplankton abundances (predominantly cyclopod copepods and cladocerans characterized by an overwintering benthic diapause stage) were observed in summers of years having cool surface-layer temperatures in the early summer. To George and Harris, these observations suggest that zooplankton abundance differences are caused by phytoplankton production differences (with cool years being more productive).

An alternate interpretation of the zooplankton/temperature coherence involves a direct link between the zooplankton and the physics. Unusually high wind mixing in the early summer would generate cooler surface waters but warmer bottom waters, which may be conducive to increased incidence of metamorphosis from the diapause stage. The food-chain hypothesis (wind mixing, phytoplankton production, zooplankton abundance) based on the match/mismatch theory is not the only interpretation of the empirical observations. In our view the match/mismatch theory, an elegant development of Hjort's critical-period hypothesis that has stimulated considerable interesting research, cannot be substantiated on the basis of the literature to date.

Concluding Remarks

In this chapter disparate aspects of the recent literature on the population regulation of marine fish have been discussed, with a heavy emphasis on recent work on Atlantic herring. Populations of marine fish having characteristic scales of abundance differing over several orders of magnitude have been observed to persist in very precise geographic locations over several decades. In some cases spawning populations have been observed for more than a century. This persistence and stability in population characteristics have occurred in spite of a high and continuing level of commercial exploitation. In our reading of the literature the present theoretical or conceptual models do not adequately account for the empirical observations. Some characteristics of the population biology are simply not addressed in the recent conceptual literature. In this category consider the species-specific differences in the geographic pattern of populations and the characteristic of population richness. Also, the causes of marked differences in absolute abundance between populations of the same species are not addressed in present theory. The characteristics of populations that are addressed, in particular the control of interannual variability in year-class size, are poorly understood. This lack of explanatory power is not due to lack of effort. The recruitment variability problem in marine fish has been a hard one to solve.

The herring hypothesis addresses some of these distributional questions. Spatial constraints are considered to be important to the diverse aspects of the population regulation problem.

The material in Chapter 4, based on the approach taken with Atlantic herring, attempts to provide a new perspective on population regulation in the oceans. The approach is, in a certain sense, a retreat to the 1930s, picking up the simpler questions about populations that were addressed by Heincke, Hjort, Damas, Fulton, and Schmidt (as summarized in Chapter 2). Why do fish spawn where they do? What are the self-sustaining populations of a particular species? Why are some species (such as European eel) panmictic and other species population rich (such as Atlantic herring)? Why in fact are most species composed of self-sustaining populations?

The starting point, then, is the population systematics literature that is well summarized for all animal groups by Goldschmidt (1939, chap. 3), Mayr (1942, chap. 6), and Rensch (1959, chap. 3). It was this literature on populations that led to the *Rassenkreis* and the biological species concept. It was also the substantive contribution to the evolutionary synthesis by the systematists and natural historians.

Given the increased empirical information base in the marine fisheries biology literature, it may be fruitful to re-ask some of these older questions. The population-level characteristics of abundance and temporal variability are considered ancillary to the species-level characteristics of population richness and their geographic pattern.

···4···

THE MEMBER/VAGRANT HYPOTHESIS

In this chapter we introduce a hypothesis, which we call the *member/vagrant* hypothesis, to explain and describe what regulates populations of sexually reproducing marine species. As background for the new hypothesis, however, let us first consider why sexual reproduction is the dominant mode underlying population persistence, and thus why or in what sense free-crossing (or outcrossing) is a constraint.

Constraint of Sexual Reproduction

In the translation of Chetverikov's work, *free-crossing* was used to indicate the mixing, or "crossing," of genetic material between members of a population during sexual reproduction. In the more recent literature, *outcrossing* has been used for the same concept. Chetverikov (1926) argues that in sexually reproducing species free-crossing is a conservative function that is balanced or opposed in evolution by natural selection. He states, for example (p. 181):

> In the foregoing analysis of free-crossing, we tried to establish its role as a factor stabilizing a given population. In its very essence it is a conservative factor, preserving the genotypic composition of the species in the condition in which it is found at a given moment.
>
> Natural selection . . . is . . . its direct antagonist. If free-crossing stabilizes the population, then selection, on the contrary, all the time displaces the equilibrium state, and, if in this sense we may call free-crossing a conservative principle, then selection, undoubtedly, is the dynamic principle, leading ceaselessly to modification of the species.

In the development of ecology much of the emphasis has been placed on an elaboration of the ecological implications of the above "dynamic principle" (natural selection through competition for limited resources). Considerably less weight has been given to the ecological implications of the "conservative principle" of free-crossing, which we consider to be an essential ecological constraint in sexually reproducing species. We will develop a hypothesis that regulation of population numbers in the oceans is predominantly a function of the constraint of free-crossing, which does not necessarily involve competition for resources. The generally accepted explanation of the selective value of sex is that it generates variability on which natural selection can act. This interpretation was promoted by Weismann in 1886. He states, for example (as quoted in Mayr 1982a, p. 705), that during fertilization

> two groups of hereditary tendencies are, as it were, combined. I regard this combination as the cause of hereditary individual characters, and I believe that the reproduction of such characters is the true significance of sexual reproduction. The object of this process is to create those individual differences which form the material out of which natural selection produces new species.

Weismann's point of view, although it has not been without critics, is still being championed. Nevertheless, Mayr (1982a, p. 599) lists the evolutionary significance of sex as one of the unresolved issues in natural selection.

An alternative interpretation of the evolutionary significance of sex, which is consistent with Chetverikov's characterization that free-crossing is a conservative principle, has been hypothesized by Dougherty (1955), Thompson (1976, 1977), and Bernstein et al. (1984, 1985). Bernstein et al. (1985) call this interpretation the *repair* hypothesis. As they explain it (p. 1277):

> Damage appears to be a fundamental problem for living systems . . . and we have argued that repair of genetic damage in conjunction with the costs of redundancy are the selective forces that lead to the origin of sex. We now propose that repair and complementation are the selective forces that maintain sex.

Two aspects of sexual reproduction are important in the repair function: recombination during meiosis (within an individual), and outcrossing (or free-crossing between individuals, in Chetverikov's terminology) during pairing. Genetic damage selects for recombination, whereas mutation (in the presence of recombination) selects for outcrossing. Bernstein and co-workers argue that outcrossing is the predominant mode of sexual reproduction in complex organisms because in diploidy there is complementation which involves the masking of recessive genes. As they state in an earlier paper (Bernstein et al. 1984, p. 336), "The diploid stage of the life cycle became the dominant stage, since diploidy allows deleterious mutations in one genome to be masked by complementary information in the other."

These arguments lead Bernstein and his co-workers to conclude that individual variation is produced as a by-product of the repair function selected for at a lower hierarchical level (those involving meiosis and diploidy). The overall hypothesis is well summarized in their concluding remarks (Bernstein et al. 1985, p. 1281):

> There are two intrinsic problems in replicating genetic information: DNA damage and mutation. We have argued that the two principal features of sex, recombination and outcrossing, are maintained, respectively, by the advantage of repairing damage and masking mutations. Variation is produced as a by-product.

The extant hypothesis (that first promoted by Weismann) in essence argues the converse, that the variation itself is selected for.

What significance does the selective value of sexual reproduction have for marine population regulation? It is perhaps important to understand the biological species concept itself. The essence of this concept is twofold. Species are defined in relation to sexual reproduction, and species consist of groups of populations (which are themselves defined in relation to free-crossing). Sexual reproduction is central to both the species concept and the definition of populations. The ultimate reasons for sex, then, are central to an understanding of population regulation in the broad sense used in this essay (richness, pattern, abundance, and variability).

Two inferences concerning populations can be inferred from the repair hypothesis:

- Due to the constraint of free-crossing in sexually reproducing species in the oceans, the complex life histories that are frequently observed ensure temporal persistence of populations in relatively fixed geographic space.

- Populations can only exist in those geographic locations within which there can be continuity in the life cycle; i.e., in a geographical setting within which retention (membership) exceeds losses (vagrancy) in some integrated sense for the life cycle as a whole.

The evolution of sexual reproduction in marine organisms, with the associated constraint of free-crossing of genetic material, introduced a new mode of selection (Bernstein et al. 1984). The primary challenge to persistence of a population of a given marine species became not just survival of the individual to maturity, as is the case for asexual species, but the ability to find a mate having a similar genome at this time.

To introduce the implications of this new constraint and its impact on population ecology, we first discuss species that complete their life cycle as plankton, because they are easier to conceptualize. The concepts can then be enlarged to include life cycles involving either a nekton phase (fish and pelagic invertebrate species) or a relatively fixed phase at the sea floor interphase (diverse benthic species).

Because the ocean environment is highly dispersive and, from the animals' perspective, dilute, it is not difficult to envision that with time subsequent to birth the chance that one individual will encounter another with similar genetic material decreases monotonically. Diffusion itself from a point source for nonmobile drifting organisms, or random movement for mobile organisms, minimizes the frequency of sexual encounter that is necessary to allow persistence of the population. A new phenomenon in the evolution of life thus was generated coincident with the emergence of sex: the constraint of physical relationship between individuals of populations to ensure persistence. More fundamentally, the very existence of populations as defined in this essay in relation to the shared gene pool emerged with the evolution of the sexual mode of reproduction. In this interpretation of the evolution of life, the advantages of masking chemically generated variability in the genetic material (through the sexual mode of reproduction) was gained at the cost of freedom of the individual in geographic space. Space relationships with others of a similar genome suddenly became a constraint. Species, as a phenomenon, emerged.

Clearly, any behavior by an animal will be highly selected for if it enhances retention of the individuals of the population in relatively fixed geographic space in proximity to the source of reproduction itself, and thus increases the probability of sexual encounter. Survival itself is not the only issue; finding a mate in a diffuse environment at dilute concentrations becomes the additional, perhaps more critical, challenge. The animal that outcompetes others for resources may survive longer; but if a mate is not encountered, obviously nothing in the genome can be selected for. Also, populations will not persist in areas where retention is not sufficient to generate appropriate rates of encounter of like genomes within the time scale of the life history. In this way, it is argued, coincident with the evolution of sexual reproduction in the oceans, the features of the physical geography in relation to diffusion-retention became critical to the persistence of plankton populations, and the constraint of sex generated persistent geographic patterns in populations in the oceans.

In sum, the bifurcation from asexual to sexual reproduction generated several new phenomena: (1) a mode of selection that ensures physical encounter at maturity; (2) behavior between individuals of similar genomes; (3) persistent population patterns in geographic space; and thus (4) the biological species itself, or the *Rassenkreis* of Rensch. As a result of the new mode of selection (that involving ensurance of physical encounter at maturity), life cycles of sexually reproducing animals are considered to be primarily mechanisms to ensure persistence of populations in specific geographic space (this will be discussed further in the concluding chapters of the essay). The second phenomenon involves the diverse behavioral characteristics of animals in relation to what Paterson (1982) has usefully defined as the Specific Mate Recognition System (SMRS). The third and fourth phenomena, the persistence in population patterns on an ecological time scale, are startling. In a very real sense it can be said that following the evolution of sex, order (in this case the generation of persistent geographic patterns in population structure) was created out of chaos (the inferred, continually changing geographical patterns that are generated by asexual reproduction).

The Hypothesis Defined

The member/vagrant hypothesis comprises three statements:

1. Population *pattern* and *richness* are functions of the number and location of geographic settings (within the overall distributional area of the species) within which the species' life cycle is capable of closure.

2. *Absolute abundance* is scaled according to the size of the geographic area in which there is closure of the life cycle of the free-crossing population. Abundance may ultimately be spatially defined in relation to the size of the spawning and early life-history distributional areas, which are themselves defined in relation to life-cycle continuity.

3. *Temporal variability* in abundance is a function of the intergenerational losses of individuals (vagrancy and mortality) from the distributional area that will ensure membership within a given population. For populations with a planktonic stage of the life cycle, physical oceanographic processes may dominate in generating the variability in losses from the distributional area.

The hypothesis emphasizes that membership in a population in the oceans requires being in the right place at the right time in the life cycle. It implies that animals can be lost from their population and become vagrants—or expatriates, to use a term from the zooplankton literature. Life cycles are considered mechanisms of achieving closure within particular geographical settings which impose spatial constraints.

This concept of complex life histories of populations is represented schematically in Figure 4.1. At any stage of the life cycle an individual can become separated from its population. The area for successful reproduction is considered to be geographically constrained in relation to the life cycle as a whole. In this illustration, sexual reproduction at other geographic locations along the central line does not allow the progeny to recruit to the population. The hole in the line represents in a single dimension a geographic area of three dimensions that is appropriate for spawning by a particular population.

A population can lose members either by *spatial* or by *energetics* processes. We define spatial processes to include geographic displacements from a distributional area appropriate for continued mem-

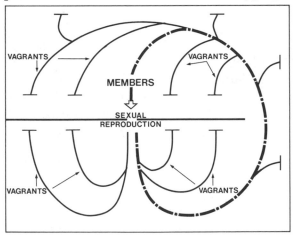

Fig. 4.1 Life-cycle closure of a marine population with a complex life history in relation to spatial constraints. It is inferred that there are chances of becoming a vagrant at different phases of the life cycle (see text for discussion).

bership in the population. Spatial processes do not necessarily involve mortality; however, they do include inability to find a mate who is a member of the population, as well as sexual reproduction in inappropriate geographic locations for continued membership by the progeny. In energetics processes we include predation, disease, and starvation (and their interactions). Such losses from a sexually reproducing marine population with a complex life history are listed in Table 4.1.

It is perhaps useful in a conceptual sense to identify separately these two categories of losses from sexually reproducing populations. Spatial losses do not involve competition for food resources or avoidance of predators. Yet they can be density dependent. In this sense the carrying capacity of a population in the marine environment could be defined below the level of food limitation and without predator control. Biological interactions involving food or predation are not required for the definition of absolute abundance (i.e., carrying capacity), or for stability and persistence. This conclusion is counterintuitive if population patterns are not considered in relation to geography. Once life cycles are associated with particular features of the physical geography, however, the argument is a simple one. The losses from a population because of spatial processes are indicated by an asterisk in Table 4.1.

It is inferred that for marine species with complex life histories

Table 4.1 Some spatial and energetics loss processes in the life cycle of a sexually reproducing species. Spatial processes are indicated by asterisks.

A.* Egg and larval losses due directly to physical processes of diffusion and advection (i.e., losses from distributional areas of the population).

B. Egg and larval mortality caused by predation, disease, and starvation.

C.* Juvenile losses from the population due directly to spatial constraints (juvenile vagrants).

D. Juvenile mortality caused by predation, disease, and starvation.

E.* Adult losses from the population due directly to spatial constraints (adult vagrants).

F. Adult mortality caused by predation, disease, and starvation.

G.* Adults which do not secure a mate or which reproduce in the wrong place for life-cycle continuity (although they are not vagrants) due to spatial constraints.

H. Adult reduced fecundity due to food limitation.

I.* Offspring which abort or are infertile due to biochemical (spatial) constraints.

the proportion of the total losses from the population due to spatial processes may be much larger than losses due to energetics processes. Further, it is noted that A* + C* + E* + G* + I* can be sufficient for population regulation. There is no requirement that abundance of the population be limited by density dependence on food availability or predation. Losses due to spatial processes could be sufficient for the regulation of absolute abundance and temporal variability. For this to be the case, all that is required is that there be some density-dependent losses associated with the spatial processes themselves. Vagrancy has to be density dependent on occasion. The existence in principle of this sort of control is illustrated in Figure 4.1 by the limited space for successful sexual reproduction.

The point being made here is an extreme one. We are not suggesting that food limitation, predation, and disease are not important in the oceans. Rather, we argue that such so-called energetics processes (acting in a density-dependent manner) are not required for population regulation. All of the density dependence could be generated by spatial processes alone. Further, a large component of population losses due to energetics processes may be density independent, just as a proportion of the losses due to spatial processes may be density dependent (this point is illustrated in Figure 4.2). In short, both categories of losses from populations can be either density independent or density dependent.

Margalef (1985), addressing the incompleteness of the equations of population growth, states (p. 202), "Volterra-Lotka models have various disadvantages: they ignore the discontinuity of individuals

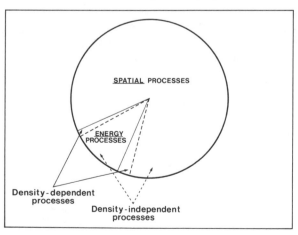

Fig. 4.2 Proportional losses (vagrancy and mortality) due to spatial and energetic processes. Each of these two defined categories of losses from a population may involve density dependence and density independence. In this representation the losses due to spatial processes predominate.

(quantification) and *space*, and operate in a sort of 'ether' of very unreal properties" (emphasis added). Leaving out the constraint of discontinuity of individuals of sexually reproducing species (individuals are members of particular populations which require being at the right place at the right time) and space may have narrowed the focus on population regulation questions in an inhibiting manner. Further, the two types of processes involved in population regulation (i.e., energetics and spatial processes) may be involved in separate aspects of evolution. In this conceptualization the energetics, or food-chain, processes lead to intraspecific competition for resources. Such competition generates selection for a certain class of adaptations associated with food-chain events. The spatial processes generated by the constraints of free-crossing in sexually reproducing animals result in relational events (membership, vagrancy), which generate what can be defined as life-cycle selection. Life-cycle selection may be considered to include sexual selection. Darwin (1859, p. 88) defines sexual selection as being a process, separate from natural selection, that does not involve the struggle for existence: "This depends, not on the struggle for existence, but on a struggle between the males for possession of the females; the result is not death to the unsuccessful competitor, but few or no offspring."

In the same sense, life-cycle selection involves a metaphorical struggle to maintain membership in the population but does not necessarily involve mortality at the time of vagrancy. Life-cycle selection does not involve intraspecific competition for limited food resources or the avoidance of predators. These contemporaneous population regulation processes, and their association with so-called energetics adaptation and speciation, are summarized in Figure 4.3.

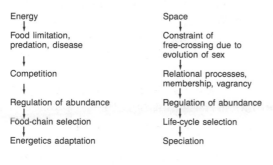

Contemporaneous Population Regulation Processes

Energy
↓
Food limitation, predation, disease
↓
Competition
↓
Regulation of abundance
↓
Food-chain selection
↓
Energetics adaptation

Space
↓
Constraint of free-crossing due to evolution of sex
↓
Relational processes, membership, vagrancy
↓
Regulation of abundance
↓
Life-cycle selection
↓
Speciation

Fig. 4.3 Inferred effects of energy and space constraints on population regulation of sexually reproducing species in the oceans. It is proposed that the two defined components of population regulation are part of the mechanisms of decoupled processes of energetics adaptation and speciation.

The evolutionary implications of the member/vagrant population regulation hypothesis are developed further in the latter half of this essay.

A recent study by Fleminger (1985) on sex change in a copepod population at certain times of the year illustrates how important spatial processes can be. The population in question persists in an upwelling zone off the coast of northern California. The life cycle involves a diapause phase in deeper water at the late copepodite stage, presumably to ensure persistence of the population in particular geographic space. In midwinter, following diapause, the copepodites migrate to the surface layer and metamorphose to adults. Because of the difference in growth or maturation rates between sexes, few mature females are available for reproduction in midwinter. Fleminger, from a detailed analysis of the morphology of the male and female antennae, concludes that at this time of the year a proportion of the males change sex, becoming what he calls "quadritheks." "At this time of few females," Fleminger states, "quadritheks have a greater likelihood of contributing to the next generation than early males, many of whom probably die without ever mating."

This latter point forcefully emphasizes the constraint of sexual reproduction on population regulation. Many males in midwinter, although presumably good competitors for food and avoiders of predators, will not contribute to the population because they are in the right place at the wrong time. The alternative pitfall, being in the wrong place at the right time for contributing to the population, is easy to visualize. The term *expatriate* is used frequently to describe zooplankton that will not be contributing to the next generation of their source population. Many of these male copepods off California will perhaps die of old age before successful mating. It is suggested that *life-cycle selection* may be a useful term to describe selection processes generated by population losses that do not involve intraspecific competition for limited food resources or predator avoidance. Sex change in midwinter by the quadritheks may be an example.

New terminology in ecology should not be defined without an obvious need. It will be argued in the concluding three chapters of the essay that a modest reconceptualization of the regulation of abundance question (as stated in the member/vagrant hypothesis) provides increased explanatory power for some contentious issues in evolutionary biology. Separating natural selection into two ecologically distinct processes (i.e., food chain and life-cycle selection) is useful towards that end. The ecological phenomena, or constraints, are distinct; and their influences are argued to be separable. Natural selection in its original definition is associated strongly with intraspecific competition for food resources and avoidance of predators. Thus, food-chain selection is roughly synonymous with natural selection as defined within Darwinism.

Natural selection, however, has taken on a different meaning within the evolutionary synthesis. The ecological component of natural selection has faded into the background, whereas the population genetics component (i.e., natural selection as changes in frequency distribution in particular traits through time irrespective of the selection process) has come to the foreground. In modern usage, sexual selection is a component of natural selection. This contrasts to Darwin's original treatment (as indicated in the quotation on page 75). Rather than restrict the definition to its original sense in discussing the implications of the member/vagrant hypothesis, we introduce two ecological components of the broadly defined natural selection concept: food chain and life cycle.

The conclusion that physics may predominate over food-chain processes in the population biology of marine species is not new. McGowan and Brinton (1985), referring particularly to plankton studies in the Calfornia Current, make this point:

> Thus the discovery of a vast amount of stirring and mixing of populations led to the speculation that much population biology and community diversity here was to be understood in terms of the physics of water movement, rather than biological functions such as food limitation, energy flow or competitive interrelationships between populations. This was such a heretical idea at the time (1954) that Johnson and his students felt more data, particularly descriptive data, were required before publishing it.

In the next four chapters some of the accumulated descriptive studies carried out predominantly since the 1960s are reviewed in relation to the member/vagrant hypothesis.

MARINE FISH POPULATIONS

Observations selected from the accumulated fisheries biology literature provide some support for the member/vagrant hypothesis, which incorporates the critical role of physical geography in ensuring sufficient retention at the early life-history stages to permit life-cycle continuity and thus population persistence. It is recognized that the nature of the argument being developed here under the rubric "support" is very much a consistency argument rather than direct evidence of the role of physics in population regulation.

Population Richness

We may take it as fact that the population structure and richness of many marine fish species (i.e., the relative degree of population richness) are determined by constraints at the early life-history stages. This is well recognized for certain anadromous species such as Atlantic salmon (e.g., see Harden Jones 1968 and Hasler et al. 1978) and American shad (Leggett and Whitney 1972; Carscadden and Leggett 1975; Dadswell et al. 1983; Melvin et al. 1985). In those two species, the population structure is a function of the number of rivers flowing into the northern Atlantic, from all land sources for Atlantic salmon and from the eastern coast of North America for American shad.

Three well-documented sets of observations are considered important. First, both species home to specific river systems (Figure 5.1). Second, egg and larval phases are completed for both species within those river systems; that is, the early life-history stages are retained by an interaction between the behavioral characteristic of the species and the particular physical geography of the river system. Third, there is extensive mixing between populations during the juvenile and adult phases of the life histories. The discreteness between populations at the early life-history stages is easy to visualize. The point that is stressed here in relation to the subsequent between-species

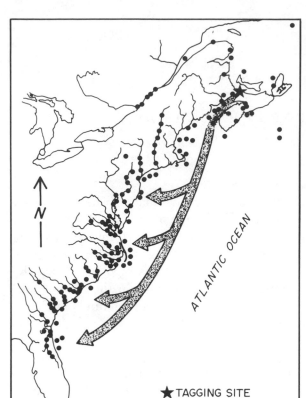

Fig. 5.1 Recapture sites of American shad returning to spawn in the rivers of their origin. (Redrawn from Dadswell et al. 1983)

comparisons is that population richness and pattern for Atlantic salmon and American shad are defined at the early life-history stages.

The anadromous rainbow smelt (*Osmerus mordax*) is somewhat less population rich than Atlantic salmon and American shad. For example, Fréchet et al. (1983) have identified only three populations of rainbow smelt in Québec waters: those spawning in rivers flowing into (1) the Bay of Chaleur, (2) the south shore of the St. Lawrence estuary, and (3) the Saguenay fjord. Those authors suggest that the Québec smelt home, not to specific natal rivers, but to one of several rivers flowing into an estuary, fjord, or coastal embayment. Ouellet and Dodson (1985) report that the larvae from one of the rivers (the Boyer) flowing into the St. Lawrence estuary are retained downstream of the natal river, in the St. Lawrence estuary, by vertical migration in combination with the two-layer circulation system.

The distributional evidence for this anadromous species again strongly suggests that population richness and pattern are defined at the early life-history stages. In this case the geographic area of retention at the early life-history stage, which is inferred to define the population structure, is in coastal embayments or estuaries rather than in the river system itself. As a result, the smelt is less population rich than the other two anadromous species considered.

The evidence supporting the conclusion that population richness is determined at the early life-history stages is not so straightforward in ocean-spawning species, due in part to the long-held concept of larval drift as first defined by Fulton (1889). In spite of this conceptual constraint on the sampling design of fish larval studies and their interpretation, many empirical observations support such a conclusion.

The evidence that the population richness and pattern of Atlantic herring (which in fact generated the hypothesis being evaluated) are determined at the early life-history stages was summarized in Chapter 3 (from Iles and Sinclair 1982; Sinclair and Iles 1985). Retention of larvae for the first few months of life has been documented for both coastal embayments or estuaries (Graham 1972, 1982; Grainger 1980; Fortier and Leggett 1982; Henri et al. 1985) as well as for open-ocean areas, by Boyar et al. (1973) and Bolz and Lough (1984) for Georges Bank and by Zijlstra (1970) for Dogger Bank. The mechanism by which discreteness is maintained (vertical migration in relation to the estuarine circulation) has been well described in estuaries (Graham 1972; Fortier and Leggett 1982; Henri et al. 1985) but only inferred in the open-ocean spawning areas. Since tidally dominated circulations are characteristic of many larval distributional areas of Atlantic herring populations, it is probable that such features play a role in facilitating retention.

Atlantic cod is also population rich. Maintenance of discrete egg and larval distributions from well-identified spawning populations has been described, for example, for Browns Bank, Sable Island Bank, Flemish Cap, and Magdalen Shallows off Canada; and for Faroe Bank, Faroe Plateau, and the Lofoten area off Europe (Table 5.1). The egg and larval distributions, however, generally have been interpreted in relation to drift and dispersal rather than persistence within particular geographic locations. For example, Hislop (1984, p. 311) has concluded

Table 5.1 Studies supporting egg and larval retention for haddock and Atlantic cod.

Study	Population
Atlantic cod	
O'Boyle et al. 1984	Browns Bank
O'Boyle et al. 1984	Sable Island Bank
Lett 1978	Magdalen Shallows
Anderson 1982	Flemish Cap
ICES 1979[a]	Faroe Plateau
ICES 1979[a]	Faroe Bank
Ellertsen et al. 1986; Wiborg 1952, 1960a, 1960b	Lofoten area
Haddock	
Smith and Morse 1985	Georges Bank
O'Boyle et al. 1984; Koslow et al. 1985	Browns Bank
O'Boyle et al. 1984	Sable Island Bank
Saville 1956; ICES 1979[b]	Faroe Plateau
ICES 1979[b]	Faroe Bank

[a]Support inferred from pp. 18 and 19 and Figs. 25 and 26.
[b]Support inferred from pp. 19 and 20 and Figs. 28 and 29.

that the high egg production of cod "ensures widespread dispersal of the spawning products."

The rich literature on Atlantic cod egg and larval distributions cannot be critically evaluated here, but the data summarized in Table 5.1 suggest that population richness and pattern of Atlantic cod are a function of constraints at the planktonic egg and larval phases of the life history. The spatial distributions of cod eggs and larvae on the Scotian Shelf off Nova Scotia at different months are illustrated in Figure 5.2. The similarity in the patterns of egg and larval distributions at any one time, as well as the persistence of the larval distributions over certain offshore banks between months, led O'Boyle et al. (1984) to conclude that cod eggs and larvae, for these areas at least, are retained in particular physical oceanographic locations. It was inferred that the gyral circulation associated with offshore banks contribute to egg and larval retention. Gagné and O'Boyle (1984), in a more detailed analysis of the ichthyoplankton data collected on the Scotian Shelf in conjunction with the groundfish survey data, strengthened this interpretation of the empirical observations:

> Our analysis of the distribution of one and two year old cod shows that they are concentrated over the banks identified as

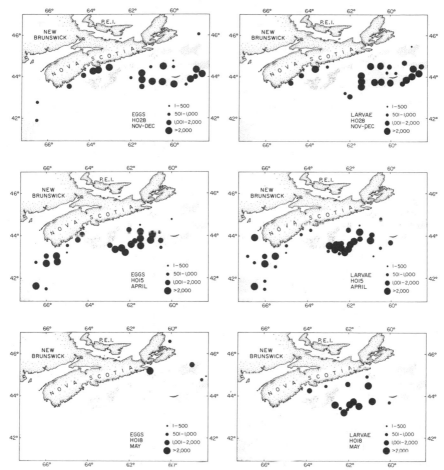

Fig. 5.2 Distribution of eggs and larvae of Atlantic cod on the Scotian Shelf during winter and spring. (Reproduced from O'Boyle et al. 1984)

major spawning areas, i.e. Banquereau, Middle and Sable Banks. This strongly suggests that the spawning and the nursery grounds are often within the same geographical area. From this evidence, we conclude that over most areas of the Scotian Shelf, cod larvae do not drift from spawning grounds to nursery grounds; instead they are generally retained within large areas where both types of grounds are located.

The observations on the distributions of cod larvae from May to August over the Flemish Cap off Newfoundland (Figure 5.3) also suggest that the planktonic early life-history stages are retained over the Flemish Cap for several months. For the two distinct populations of

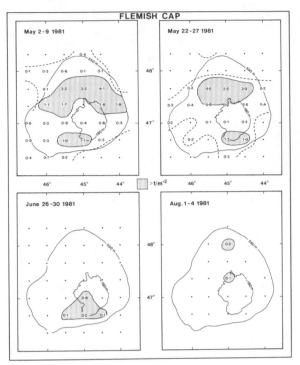

Fig. 5.3 Distribution (in numbers per cubic meter) of Atlantic cod larvae on the Flemish Cap from May to early August. (Redrawn from Anderson 1982)

cod that have been identified on the Faroe Plateau and around the Faroe Islands, spawning locations and distributions of juveniles less than a year old are discrete between the two areas (Figure 5.4). The ichthyoplankton stages of the two populations are probably retained around the Faroe Plateau and over the Faroe Islands. Evidence for an anticyclonic circulation on Faroe Bank, which should enhance egg and larval retention, is presented by Hansen et al. (1986). Even where the juvenile distributional area is not coincident with egg and larval distributional areas, such as the Lofoten and Vestfjord spawning population off northern Norway, there is evidence that eggs and larvae are retained over the offshore banks for a few months.

In the interpretation now accepted, the eggs and larvae drift passively from the spawning site to the nursery area to the north and east. An alternate interpretation for this cod population is that at the planktonic stages they are retained over the banks and within the Vestfjord for a certain period (Ellertsen et al. 1986) and then, either at the late larval stage or after metamorphosis, actively migrate to the juvenile nursery area. It is suggested here that the very existence of the self-sustaining population in this area is due to the retention charac-

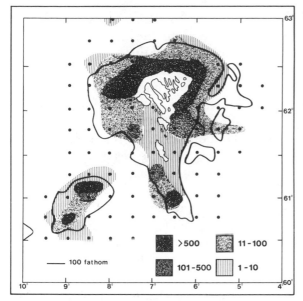

Fig. 5.4 Distribution (in numbers per standard haul) of recently metamorphosed Atlantic cod juveniles from discrete populations on Faroe Bank and around the Faroe Islands. (Reproduced from ICES 1979)

teristics of the Lofoten-Vestfjord physical oceanography. Following this logic, the precise location of spawning is defined, not by the *residual* circulation linking a spawning area to a nursery area, but by the bank *recirculation* features whereby egg and larval discreteness can be maintained for a few months.

In each of the above examples, the cod populations have been identified by their adult characteristics. More recent distributional observations by other scientists on the early life-history stages suggest that the eggs and larvae are retained in well-defined geographical areas having particular physical oceanographic properties. Mixing between populations of cod, however, does occur at the adult phase during summer feeding (Wise 1962) and overwintering (Templeman 1962).

Egg and larval retention for haddock spawning on offshore banks has been demonstrated or inferred for Georges Bank, Browns Bank, Sable Island Bank, Faroe Plateau, and Faroe Bank (Table 5.1). Saville (1956) postulates (p. 11) that retention is generated by an anticyclonic eddy system around the Faroes: "It would seem necessary to postulate such a system in any case to explain the retention of the haddock spawning products within the area and, with the possible exception of 1950, there is no evidence to suggest that there is any appreciable loss of these products by drift out of the area."

It can be inferred from detailed studies of the oceanography of

Browns and Georges banks, in the Gulf of Maine area, that larval behavior as well as circulation is critical to the persistence of the distributions on the banks. Both banks are highly dispersive environments. The dispersion time of passive particles, without considering the circulation, is 10–15 days for Browns Bank and 50–60 days for Georges Bank (P. Smith, Bedford Institute of Oceanography, personal communication). Residual gyres generated by tidal processes are also characteristic of both banks (Greenberg 1983). Drogue studies on Browns Bank at midwater depths show that the residence time, on average, is 14 days (P. Smith, personal communication).

The "event" scale of physical oceanographic phenomena (such as shelf-water entrainment into warm-core eddies and advection of shelf water seaward due to storms) has been identified as a constraint to egg and larval persistence on shelf banks in this particular area. Yet persistent discrete egg and larval haddock (and cod) are observed on both banks (Figures 5.5–5.7); and more, larger larvae are observed within the well-mixed area of Georges Bank than in the stratified water around the circumference (G. C. Laurence, National Marine Fisheries Service, personal communication). In addition, discrete aggregations

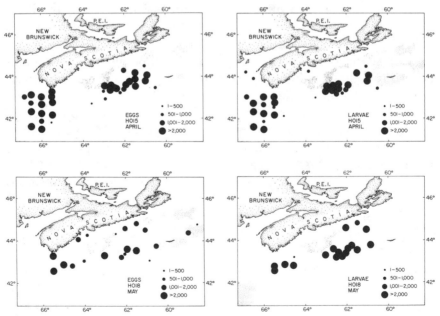

Fig. 5.5 Distribution of haddock eggs and larvae on the Scotian Shelf during the spring. (Reproduced from O'Boyle et al. 1982)

of 0-group haddock (postlarvae less than one year old) are observed on at least Georges Bank and Sable Island Bank (Scott 1982, 1984) in the summer (the situation for Browns Bank is less clear). Thus, some behavior has to be attributed to the early life history, particularly on the smaller Browns Bank, to account for the observed distributions.

It is not clear at this stage how either diurnal or ontogenetic depth changes by the larvae and postlarvae can account for the observed distributions. There is little shear in the tidal circulation with depth in the top 30 m of the water column within which most of the larvae are distributed. However, wind-generated circulation effects, which are superimposed on the tidal circulation, are depth specific and may generate sufficient vertical structure to be "used" by vertically migrating larvae to enhance retention over the natal banks.

Tagging results of haddock indicate that (1) the adult distributional areas are considerably broader than the spawning and egg and larval retention areas and (2) there is some mixing between haddock of different populations during summer feeding and overwintering (for example, McCracken 1959; Halliday and McCracken 1970). On the basis of the distributional data at various life-history stages, it is tentatively inferred here that haddock population richness (which is considerably less than that observed for herring and cod) is defined at the planktonic egg and larval phases of the life history. In the northwestern Atlantic, at least, cod populations exist in essentially every location that a haddock population spawns; but there are also many geographic locations where only cod populations can persist (e.g., the many so-

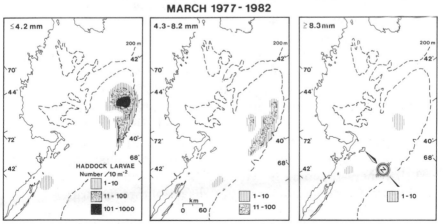

MARCH 1977 - 1982

Fig. 5.6 Distributions of three size classes of haddock larvae of the Georges Bank population during March. (Reproduced from Smith and Morse 1985)

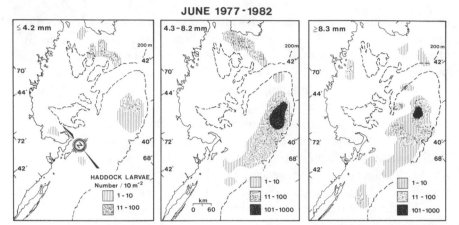

Fig. 5.7 Distributions of three size classes of haddock larvae of the Georges Bank population during June. (Reproduced from Smith and Morse 1985)

called local cod populations maintained in coastal embayments). This difference in population richness between species whose early life-history requirements are similar suggests that there may be very subtle differences in the behavior of the larvae that allow cod populations to be more numerous.

Winter flounder population structure has not been well described. There has been, however, an excellent study on the early life history of a selected well-defined population. Pearcy (1962) elegantly demonstrated the interaction between ontogenetic changes in winter flounder behavior at the early life-history stages in relation to the estuarine circulation that permits maintenance of a discrete early life-history distribution within the Mystic River estuary. From tagging experiments it has been concluded that winter flounder are rather stationary and that the species is population rich (Perlmutter 1947). Saila (1961) showed that winter flounder return to the same area each year to spawn. It is thus not possible to conclude for winter flounder that the population structure itself is a constraint of requirements at the egg and larval phases (since all phases are fairly discrete from contiguous populations); but the behavioral features identified at the early life history, which are linked to physical oceanographic features, suggest that they are aiding retention of the population in specific geographic areas rather than permitting dispersal through drift.

Similar evidence on egg and larval retention in relation to population patterns can be inferred for yellowtail flounder (*Limanda ferruginea*). Studies on juvenile and adult characteristics and migrations (Lux 1963) have led to the conclusion that yellowtail flounder are generally sedentary and that separate populations are sustained in close proximity to each other. For example, separate populations of this species are observed on Georges Bank, from Nantucket Shoals to Long Island, and from Cape Cod to Cape Ann. Smith et al. (1978) describe vertical distribution of yellowtail larvae and ontogenetic changes in behavior for the Middle Atlantic Bight population. They indicate that vertical movements and feeding are not directly related. The younger larvae stay below the thermocline, but intermediate-size larvae move through it and can be found throughout the water column. Larvae greater than 10 mm apparently spend some time on the bottom. There is little dispersal at the larval stage due to the surface-layer and wind-driven currents. Smith et al. (1978) state,

> Our conclusion is supported by Royce et al. (1959). Similarities in patterns of distribution between eggs and larvae led them to conclude that larvae were demersal before much horizontal drift occurred. It seems worth noting that the smallest larvae, those least able to swim with directed movements, did not ascend to the surface at night. They remained below the shallow thermal gradient, where they were unaffected by wind-driven circulation.

On the basis of this limited evidence on yellowtail larval distributions and the more extensive literature on the population patterns of yellowtail flounder, it is again argued that the richness and geographic pattern observed are a function of events at the early life history.

Atlantic mackerel (*Scomber scombrus*) population structure was initially described by Garstang (1899) and Williamson (1900) in the eastern Atlantic and Sette (1943) in the western Atlantic. Subsequent studies have not markedly changed their major conclusions. The Atlantic mackerel is not population rich, consisting of only two large populations in the western Atlantic and perhaps two or three in the eastern Atlantic. Large-scale migrations are characteristic of this species, and during the overwintering adult phase there is mixing between populations. The western Atlantic contains two extensive but definable egg and larval distributional areas: the southern Gulf of St. Lawrence (Lett 1978) and the Middle Atlantic Bight (Berrien 1978). The distribution of mackerel eggs in the Middle Atlantic Bight is shown in Figure 5.8 and

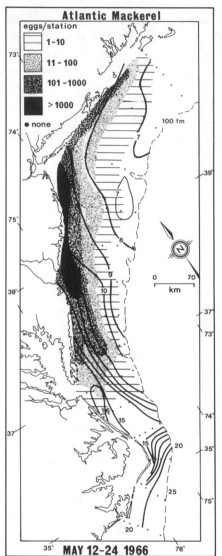

Fig. 5.8 Distribution of Atlantic mackerel eggs in the Middle Atlantic Bight. (Reproduced from Berrien 1978)

of larvae in the North Sea in Figure 5.9. The egg and larval retention areas for Atlantic mackerel as a whole are illustrated in Figure 5.10.

It is proposed here that the egg and larval retention areas used by this species involve larger-scale circulation features, of which there are not many within its distributional limits. As a result, only two populations can be sustained, for example, in the western Atlantic. Again, the population structure of Atlantic mackerel is inferred here to

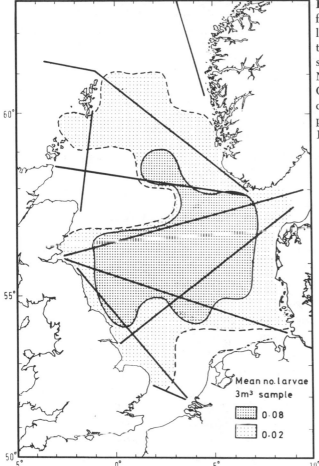

Fig. 5.9 Distribution from 1958 to 1968 of Atlantic mackerel larvae in the North Sea. The solid straight lines across the North Sea represent the Continuous Plankton Recorder routes. (Reproduced from Johnson 1977)

Mean no. larvae 3m³ sample

0·08
0·02

be defined at the early life-history stages (in this case in relation to large physical oceanographic current systems). The discreteness of the egg and larval distributions in the northeastern Atlantic in relation to either population structure or well-defined physical oceanographic features is not well understood (see Hamre 1980 and Johnson 1977 for reviews).

Shortfin squid and Atlantic menhaden are population poor. For each species the distributional limits of the species itself define the population. Clearly, however, the physical oceanographic systems used by these species at the early life-history egg and larval phases are in each case unique. It is hypothesized by O'Dor (1981), Rowell et al. (1984), and Trites (1983) that shortfin squid use the interphase between the Gulf Stream and the slope water as the larval distributional area

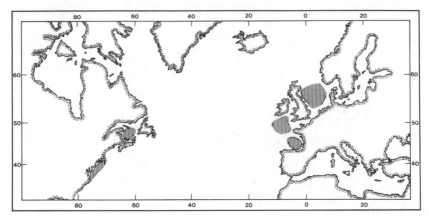

Fig. 5.10   Larval retention areas for Atlantic mackerel, a population-poor species.

(there is only one such feature). Atlantic menhaden use the continental shelf from Cape Cod to northern Florida as an egg and larval distributional area (see Kendall and Reintjes 1975 for a review).

The final example chosen is the European eel because it has been so extensively studied and, perhaps incorrectly, considered to be an anomaly with respect to its population structure. The European eel, like the shortfin squid and the Atlantic menhaden, is population poor. There are disagreements in the literature concerning the systematics of eels in the northern Atlantic (Williams and Koehn 1984). This uncertainty does not, however, significantly influence the conclusions drawn here. As indicated in Chapter 2, Schmidt (1930) observed no differences in meristics of European eel from his extensive sampling of rivers throughout Europe. These observations were anomalous to his observations on herring, cod, and *Zoarces* sp., and there have been ingenious attempts to account for the homogeneity of the observations (for example, Wynne-Edwards 1962). In the present perspective, however, the European eel results are clearly not anomalous, but rather part of an ordered sequence of population richness. Also, the underlying causes of the lack of population richness are plausibly interpretable.

For this species the egg and larval retention area is the North Atlantic Gyre itself (Figure 5.11), and there is only one such physical feature within the distributional range of the species (here we consider the European and American eels to be separate species). The distributional details of the early life history and the observations on homogeneity in the meristics of adult eels are described by Schmidt (1922). His conclusions about the European eel have recently been questioned by

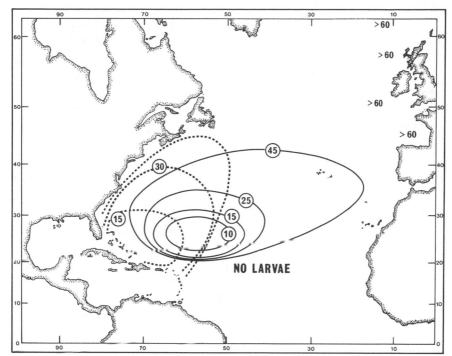

Fig. 5.11 Distribution of larvae of European eel (solid lines) and American eel (dotted lines). (Reproduced from Schmidt 1922)

Boëtius and Harding (1985) and Harding (1985). Although the published data by Schmidt indicate remarkable uniformity in meristics between eels sampled from a wide range of rivers throughout Europe, the unpublished data and some subsequent work by the authors demontrate less homogeneity. In addition, the larval distribution data in the North Atlantic Gyre suggest that the spawning location is more extensive than what Schmidt hypothesized. The subsequent research on the American eel, however, has tended to confirm the conclusion of panmixis for eels (Williams and Koehn 1984; Kleckner and McCleave 1985; references therein).

At this stage in the debate the general conclusions of Schmidt on panmixis and a single larval distributional area for each of the two eel species in the northern Atlantic has been accepted. This interpretation is supported in a recent review by McCleave et al. (in press). It is noted, however, that the strength of the conclusions has been weakened for the European eel (but not the American eel) by the analyses of Boëtius and Harding. Future studies will undoubtedly clarify this question.

In sum, there is a continuum in population richness from, for example, Atlantic salmon to European eel. In this comparative analysis it is clear that in general terms the degree of population richness is a function of constraints at the early life-history phases involving well-defined physical oceanographic or geographic features (rivers, estuaries, coastal embayments, tidal-induced circulation features, banks, coastal current systems, major ocean currents, oceanic-scale gyres). When there are many such physical features (to which the early life history is behaviorally adapted to permit retention) within the distributional range of the species, the species is population rich. With few appropriate physical features, the species is population poor. The extremes in the continuum forcefully and simply demonstrate that it is the early life-history discreteness itself that defines the population richness.

The life cycles of Atlantic salmon and European eel are almost mirror images of each other, and for these two species population richness is generated by the early life-history discreteness rather than by the juvenile and nonreproductive parts of the adult phases. There is extensive mixing between populations at the nonreproductive adult phase for salmon, and isolation of parts of a panmictic population at this part of the adult phase of eel. It is the continuity/discontinuity at the early life-history phase that leads to population richness. There is, in addition, sufficient evidence to join these extreme examples and thus describe the interspecies patterns in population richness as a continuum (Figure 5.12). Discontinuity at the early life-history stages in relation to particular oceanographic features, in conjunction with the ability of adults to home to these features for reproduction, generates the observed differences in population richness.

This summary of species-specific differences in population

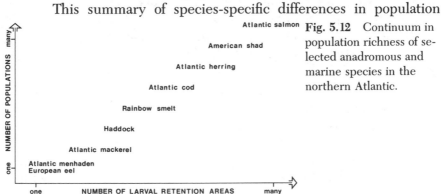

Fig. 5.12 Continuum in population richness of selected anadromous and marine species in the northern Atlantic.

richness is in no way exhaustive, but rather a taste of the overall literature. The conclusion, however, that population richness is defined at the early life-history stages to retention-diffusion processes is well supported. Burton and Feldman (1982), from a consideration of a different component of the marine literature (that on the population genetics of estuarine and coastal zone invertebrates), have also concluded that processes during the pelagic early life-history stages influence population structure.

The member/vagrant hypothesis accounts for the difference in population richness between species in strictly ecological terms. The evidence from the marine fish literature is, in our view, convincing. The empirical observations on population richness and their interpretation suggest that this process has little to do with speciation. Checkerspot butterfly and Atlantic herring are probably not in the process of speciation to a greater degree than the satyrnine butterfly and European eel. It is suggested here that differences in population richness are a function of differences in the number of geographic settings that allow life-cycle closure. In this sense descriptions of population pattern and richness are not necessarily evidence of speciation in action. This interpretation is consistent with the conclusion of Goldschmidt (1940, p. 396):

> Microevolution, especially geographic variation, adapts the species to the different conditions existing in the available range of distribution. Microevolution does not lead beyond the confines of the species, and the typical products of microevolution, the geographic races, are not incipient species.

We will return to a fuller discussion of this aspect of population regulation in later chapters. For now, it is enough to say that the interspecific differences in these simple geographic patterns and richness of populations are not generated by geographic barriers within the distributional limits of the species. Members of different populations can share a common distributional area for much of their life cycle yet sustain reproductive isolation at the time of spawning. Richness is rather a function of the number of oceanographic features that can sustain members at the early life-history stages given the particular behavioral characteristics of the species.

Absolute Abundance

The third feature of populations addressed by the member/vagrant hypothesis is absolute abundance. In Chapter 3 some evidence presented for Atlantic herring indicated that abundance is at least partially a function of the geographic scale of the physical oceanographic feature associated with spawning and early life-history distributions. In qualitative terms the same conclusion can be drawn for other marine fish species (Atlantic cod, haddock, and yellowtail flounder). Absolute abundance of populations appears to be partially a function of the geographic scale of the spawning and larval retention area rather than the geographic range at the other stages of the life cycle. Populations of marine fish are frequently observed to be mixed during the nonspawning periods (i.e., during the juvenile phase, adult summer-feeding, and overwintering). In this sense populations of different orders of magnitude may share a common distributional area at the same time; this is well recognized for anadromous species such as Atlantic salmon and American shad, but less so for marine species such as Atlantic mackerel, Atlantic herring, and Atlantic cod.

This simple observation is of interest. Fish feeding or overwintering side by side can be members of different populations. Some of the fish may be at the high end of their abundance scales, others at the low end of theirs; yet they are all sharing the same feeding and predator environment. It is difficult to envision how food limitation or predator control could operate selectively between populations. Evidence for the hypothesis relative to the definition of absolute abundance is limited and largely of a qualitative nature except for the Atlantic herring observations (Figure 3.6).

Additional support from the fisheries biology literature for the member/vagrant hypothesis deals with the variability in absolute abundance of fish populations. Several studies have inferred that interannual variability in abundance in year-class strengths is due *directly* to physical oceanographic processes that cannot lead to match/mismatch but can only be interpreted by advection. These studies conclude that the variance in year-class strengths is a function of variability in the retention of larvae in the appropriate distributional area due to advective and diffusive losses. A wide variety of fish species have been considered: Atlantic herring (Iles and Sinclair 1982), Atlantic menhaden (Nelson et al. 1977), Atlantic croaker (Norcross et al. 1984), Black Sea turbot (Popova 1972), Pacific mackerel (Sinclair et al. 1985c), Pacific hake (Bailey 1981), northern anchovy (Methot 1983), Pacific cod (Tyler and Westreim 1985), and spiny lobster (Pringle 1986).

The recent studies on recruitment variability of Labrador cod

(Sutcliffe et al. 1983) can be explained by variable advective losses of larvae from the appropriate distributional area rather than by the authors' "food-chain" hypothesis (Sinclair et al. 1986). It is not necessary that larval mortality actually occur, but rather that the larvae be in the appropriate geographical environment to be able to contribute to their natal population after metamorphosis. Methot (1983, p. 749) states for northern anchovy, "Adverse larval drift would not necessarily cause increased mortality during the age interval examined... but may affect the fraction of the surviving larvae which are entrained in the range of the juvenile habitat." In this interpretation by Methot, as well as in the study on Pacific mackerel by Sinclair et al. (1985c), larvae advected or diffused out of the appropriate larval distributional area of the population quite simply become lost fish which cannot contribute to the population's gene pool.

Methot's conclusion is fully consistent with the above-cited studies and is further supported by the repeated observation that larval abundances estimated late in the larval phase are not correlated with subsequent recruitment to the population (Sissenwine 1985). Rosenberg (1984) supports the inference by Iles and Sinclair (1982) that interannual variability in the diffusive characteristics of the larval retention area of a herring population influences subsequent recruitment to the population.

The combined time series observations over several decades on relative survival of Pacific mackerel eggs, larvae, and postlarvae (Age 1 abundance divided by the number of eggs spawned by the mature biomass), on plankton abundance in the California Current, and on the relative strength or transport of the Current itself, provide a rare direct opportunity to evaluate the importance of food and physics on the generation of year-class variability. Chelton et al. (1982) evaluate the use of a sea-level index (San Francisco, Los Angeles, and San Diego) as a measure of the temporal variability of the strength of the California Current. With the index the response of the California Current to El Niño-Southern Oscillation (ENSO) events is described over several decades (Figure 5.13). With the extensive data from the CalCOFI program the relationship between zooplankton abundance and the strength of the California Current is also documented. High zooplankton abundances of larger size categories are associated with periods of high transport in the California Current (i.e., periods of relatively lower sea level) (Figure 5.14).

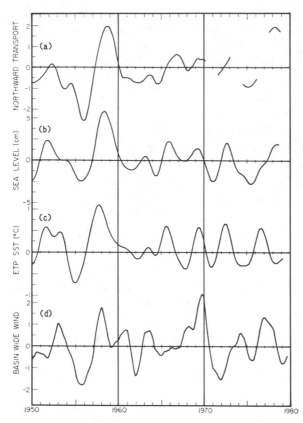

Fig. 5.13 Interannual variability in (a) transport of the California Current and (b) sea level index. Reduced southern transport during El Niño events is reflected by increased sea level. (Reproduced from Chelton et al. 1982)

There is also good information on both the relationship between large and small zooplankton (the latter being the food for many fish larvae) and the relative survival of Pacific mackerel during their first several months in the plankton. High survival rate of mackerel is associated with time periods of reduced transport in the California Current (i.e., higher sea levels) (Figure 5.15).

In sum, relatively high survival is associated with low advection and low food abundance. A component of recruitment variability can thus be attributed *directly* to the physical processes associated with ENSO events (without food-chain interactions). It is interpreted by Sinclair et al. (1985c) that reduced transport in the California Current enhances retention of the early life-history stages within the appropriate geographic area for the population. Mortality itself is not implied, but variable rates of vagrancy are, as a function of the physics.

The final source of fisheries biology support for year-class variability is on the timing of spawning and the duration of the larval phase

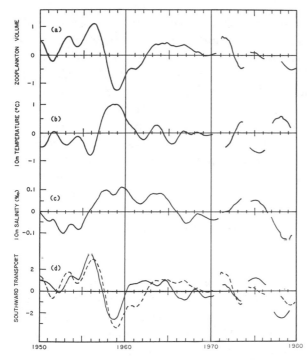

Fig. 5.14 Interannual variability in (a) estimated zooplankton abundance and (d) southward transport of the California Current. Years of strong southward flow are characterized by higher-than-average abundance of zooplankton. (Reproduced from Chelton et al. 1982)

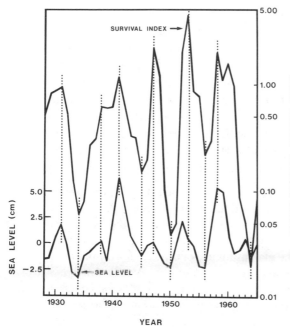

Fig. 5.15 Interannual variability in a survival index for Pacific mackerel and the sea level index of Chelton et al. (1982), which indicates the El Niño events. During El Niños (i.e., years characterized by reduced southward transport of the California Current and low zooplankton abundance), the survival of mackerel in their early life-history stages is relatively high. (Reproduced from Sinclair et al. 1985c)

itself. Recent studies of an eastern boundary current area (Parrish et al. 1981), the north temperate Atlantic (Sinclair and Tremblay 1984), and a tropical environment (Johannes 1978; Owen 1984) conclude that the spawning times of a wide range of fish species are defined in relation to the constraint of egg and larval retention in appropriate distributional areas rather than in relation to high food availability. Johnson et al. (1984) draw the same conclusion for the spawning time of blue crabs in the lower Chesapeake Bay. They suggest that hatching, which peaks in July–August, is timed to the northerly flow of surface water (p. 426) and thus provides "a mechanism for retention of larvae within the vicinity of the bay mouth, or at least to prevent their loss from the MAB [Middle Atlantic Bight] via the southward drift." These studies support the hypothesis that the physics in support of retention in particular geographic space is more of a constraint on life-history features than is food availability.

Three examples are provided for the duration of the larval phase. The larval phase of the western rock lobster (*Panulirus cygnus* George) population on the west coast of Australia lasts 9–11 months and involves current systems thousands of kilometers in extent, yet the larvae return to the appropriate juvenile distributional area along the west coast (several hundreds of kilometers in extent) (Phillips 1981). The larval phase as a whole in this example cannot be considered to be for dispersal, but rather for retention (albeit within a very large oceanographic environment) by selective advection, and thus for persistence of the population on the west coast of Australia. Ontogenetic changes in behavior at the several larval stages, in relation to the physical oceanography of the area, are interpreted to permit persistence of the population in a particular geographic location. Also, life-history events are coincident with physical phenomena permitting retention, and thus persistence, rather than with food availability. The particular time of spawning (Phillips, p. 34),

> places the phyllosoma larvae in the southeastern Indian Ocean at the time of lowest zooplankton abundance and hence food availability. One would expect that a species would evolve its biological patterns of activity (particularly those associated with reproduction) to make maximum use of times of maximum food availability. . . . That this is not the case [for] *Panulirus cygnus* presumably indicates an overriding need to achieve successful recruitment in the prevailing water circulation of the southeastern Indian Ocean.

Other species of the same genus, for example *P. interruptus* (Johnson 1960) in the California Current, have somewhat shorter larval phases (presumably because the life cycle takes place in geographically smaller oceanographic systems). The mean larval phase (15 months) of another lobster species, *Jasus edwardsii*, from New Zealand, is even longer than that for *P. cygnus* (Lesser 1978). The details of the distribution in relation to physical oceanographic processes for the New Zealand lobster are not so well described, but it may be that the between-species differences in duration of the spiny lobster larval phase are defined in relation to the spatial scale of life-cycle continuity.

The second example involving duration of larval phases is the marked between-population differences for Atlantic herring (2–11 months) (Sinclair and Tremblay 1984). This difference is interpreted to be a function of two constraints: (1) the larvae of each population are retained within a particular oceanographic environment; and (2) metamorphosis is seasonally limited. If the larval retention area is good for growth, the population can spawn in the spring and metamorphose within the acceptable seasonal envelope. If growth conditions of the retention area are less good, the larval phase has to be longer and the population must spawn earlier (with respect to the season of metamorphosis) to be able to metamorphose within the appropriate period (either in the winter or in the autumn). In this interpretation of the observations the most important constraint on the species with respect to the ability to maintain a population is the existence of a geographically stable larval retention area. If the physics is appropriate to maintain a discrete larval distribution for the first several months of the larval phase, there is sufficient plasticity in the life history of the species to stretch the duration of the larval phase (and adjust the time of spawning) to account for the particular growth constraints.

The third example is the difference in the duration of the larval phase between the American eel (one year) and the European eel (2.5 years). The growth of the American and European leptocephali is shown in Figure 5.16. The larval distributional areas were described from Schmidt's work above (Figure 5.11). The European leptocephali travel farther to return to their spawning rivers than do the American leptocephali. The difference between the two species in the duration of the larval phase appears to be a constraint of the physical geography involved in life-cycle closure, rather than the result of differences in the

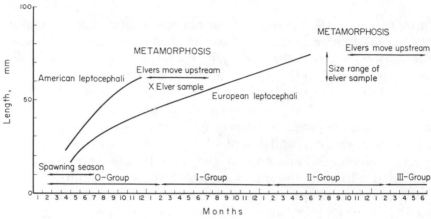

Fig. 5.16 Growth of American and European eel larvae. (Reproduced from Harden-Jones 1968, based on data from Schmidt 1922)

feeding environment. In fact the two species share parts of the same environment at this phase of the life history.

In sum, the fisheries biology literature provides some support for that part of the member/vagrant hypothesis which affirms the critical role of physical geography in ensuring sufficient egg and larval retention to permit population persistence. As stated early in the chapter, what is here called support is a consistency argument rather than direct evidence.

···6···

ESTUARINE, ISLAND, CORAL REEF, AND BENTHOS POPULATIONS

We begin this chapter by considering the estuarine literature, in particular to evaluate the ability of diverse species to maintain persistent populations in this semi-enclosed yet physically dynamic environment. For plankton species or for species with a pelagic stage in the life history, the obvious question is: how are populations sustained in the face of the strong flushing characteristics of estuaries? As a result of this unavoidable question there is in estuarine studies a much stronger focus on the retention of pelagic stages than one finds in the open-ocean literature, and behavior is considered crucial, in contrast to the inferred passive dispersal of coastal and open-ocean marine plankton.

The semi-enclosed nature of estuaries and their proximity to research laboratories and field stations have prompted numerous studies on retention of populations within estuaries, starting with those of Rogers (1940). The literature supporting the concept that separate populations can sustain themselves within individual estuaries or estuarine systems is summarized in Table 6.1. Two categories are considered: zooplankton species, and benthic species with pelagic stages of the life history. The estuarine zooplankton literature, including studies on population maintenance and flushing rates, has been reviewed by Miller (1983); and the literature on retention/dispersal of pelagic stages of benthic species in estuaries is discussed in detail by Kennedy (1982). Phytoplankton retention mechanisms in estuaries have also been noted. Anderson and Stolzenbach (1985) describe selective retention of two species of dinoflagellates in a well-mixed estuarine embayment. They state (p. 47) that "tidal mixing is very efficient and that the cells

Table 6.1 Estuarine studies describing retention of zooplankton and pelagic life-history stages of benthic species.

Study	Species	Location
Zooplankton		
Barlow 1955	*Acartia tonsa* (copepod)	Great Pond, Massachusetts
Bosch and Taylor 1970, 1973	*Podon polyphemoides* (branchipod)	Chesapeake Bay, Virginia
Jacobs 1968	*Acartia tonsa* (copepod)	Sapels Island, Georgia
Trinast 1975	*Acartia californiensis* (copepod)	Upper Newport Bay, California
Wooldridge and Erasmus 1980	*Pseudodiaptomus hussei* (copepod)	Sundays River estuary, South Africa
	Acartia longipattela (copepod)	
	Acartia natalensis (copepod)	
	Rhopalophthalamus tenanatalis(mysid)	
	Mesopodopsis slabberi (mysid)	
Benthic species		
Bousfield 1955	*Balanus improvisus* (barnacle)	Mirimichi estuary, New Brunswick
	Balanus crenatus (barnacle)	
	Balanus balanoides (barnacle)	
Carriker 1951	*Crassostrea virginica* (oyster)	New Jersey estuaries
Cronin 1982; Cronin and Forward 1982	*Rhithropanopeus harrisii* (crab)	Newport River estuary, North Carolina
Dittel and Epifanio 1982	*Uca* spp. (crab)	Delaware Bay, Delaware
	Pinnixa chaetopterana (crab)	
	Pinnixa sayana (crab)	
Sandifer 1973, 1975	*Hippolyte pleurocantha* (caridean shrimp)	York River estuary and lower Chesapeake Bay, Virginia
	Ogyrides limicola (caridean shrimp)	
	Palaemonetes spp. (caridean shrimp)	
	Hexapanopeus angustifrons (crab)	
	Neopanope sayi (crab)	
	Pinnixa chaetopterana (crab)	
	Pinnixa sayana (crab)	
	Pinnotheres maculatus (crab)	
	Pinnotheres ostreum (crab)	
	Rhithropanopeus harrisii (crab)	
	Sesarma reticulatum (crab)	
	Uca spp. (crab)	
Sloan 1985	*Lithodes aequispina* (crab)	Portland Inlet, British Columbia
Wood and Hargis 1971	*Crassostrea virginica* (oyster)	James River estuary, Virginia

would be flushed from the salt pond in the absence of a behavior-related retention mechanism."

The behavior of the pelagic stages that permits retention, and thus geographical persistence, is remarkably diverse and species specific. The details are not described here, but the ability of planktonic species, or a species with a pelagic phase of the life history, to maintain self-sustaining populations within a particular geographic space (in this case an estuary or an estuarine system) is very well documented. De Wolf (1973, 1974), in a detailed study of the larval stage of several species of barnacles, has argued (1974, p. 415) that "retention of larvae in estuaries can be explained by a mechanical process, rather than by the swimming behavior of the larvae induced or released by tide-coupled environmental factors." However, one might consider variable sinking rate itself in relation to the tidal circulation to be a behavioral trait.

It is to be noted that not all estuarine benthic species have pelagic phases that are retained in individual estuaries, or even in estuarine systems. Larvae of the blue crab (*Callinectes sapidus*), for example, an extensively studied species, are flushed out of the many estuaries along the Middle Atlantic Bight. Individual estuaries do not then have self-sustaining populations of blue crabs, but are parts of a larger geographical system which sustains life-cycle continuity (see Sulkin and Van Heukelem 1982, Epifanio and Dittel 1982, and Dittel and Epifanio 1982).

The recurrent theme in the estuarine population studies is that behavior (both daily and ontogenetic changes) in relation to physical oceanographic processes can permit retention. Figure 6.1 illustrates a well-described example of migration behavior to enhance retention within an estuary; it shows that crab larvae (stage 1 zoea and stage 4 zoea) of *Rhithropanopeus harrisii* migrate vertically as a function of the tidal cycle. For many endemic species in estuaries the mechanisms of retention have not been studied but are inferred. Miller (1983, p. 106), for example, states:

> One endemic form, however, *Eurytemora herdmani*, is distributed in a zone of the St. Lawrence Estuary centered just downstream of the null point. A population of a species of this genus is found at equivalent parts of a number of estuaries. The mechanisms by which they are retained within the estuary are obscure. The vertical center of distribution for the older life cycle stages is deep in the salt wedge. A balance between

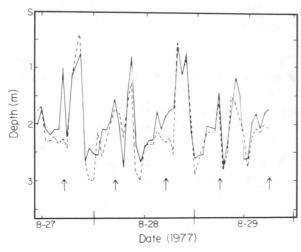

Fig. 6.1 Vertical migration of two stages of crab larvae: stage 1 zoea (solid line) and stage 4 zoea (dashed line) of *Rhithropanopeus harrisii*. Arrows indicate low tides. (Reproduced from Cronin and Forward 1982)

upstream and downstream movement is probably achieved by some form of vertical migration: ontogenetic, diel, or possibly without a timing pattern common to the whole population.

Strathmann (1982) has argued that it is retention within the natal estuary that is selected for in pelagic stages of benthic invertebrates, rather than dispersal for colonization of new habitats. This interpretation is not necessarily at odds with the studies on large-scale dispersal capabilities of larval stages of benthic animals by, for example, Scheltema (1974) and Jablonski and Lutz (1983). Retention and dispersal are two sides of the same coin. On an ecological time scale it is argued here that retention within specific geographic space to permit persistence of sexually reproducing populations is what is selected for. The unavoidable dispersal of larval stages out of the appropriate distributional area of the estuarine population is considered to be an artifact, albeit one that may be of considerable importance on an evolutionary time scale.

In sum, because estuarine species are relatively easy to study, it has been repeatedly demonstrated that both planktonic species and pelagic phases of benthic species use, with remarkable creativity, the estuarine physical processes to ensure persistence of populations in particular geographic space. The empirical evidence is compelling that many species can maintain relatively discrete populations within both stratified and unstratified estuarine circulations and also that there is considerable loss of individuals from the populations. The same sorts of phenomena are probably occurring in the open ocean, where the physical discontinuities that permit retention, and thus population persistence, are less obvious.

Oceanic islands are a second marine environment in which the persistence of populations with a pelagic phase of the life history is an obvious problem. For sexually reproducing species inhabiting the very narrow shelf areas around oceanic islands, which in some cases can be separated from other similar habitats by a thousand kilometers or more, the intuitive conclusion is that the populations of marine species around such islands are self-sustaining. If intuition is correct, there must be mechanisms by which the pelagic phases are retained. Because of this simple argument, as mentioned above for estuaries, the oceanic island marine literature places considerably more emphasis on larval retention to permit population persistence than on larval drift and dispersal. As is the case of semi-enclosed estuaries, the scientist can see the physical discontinuities (in this case widely separated oceanic islands) to which species have generated their particular population structure.

Boden (1952) and Boden and Kampa (1953) conclude that the complex circulation around Bermuda enhances retention. Emery's (1972) study of a stable eddy system in the downstream wake of the Barbados provides further support for the physical basis of retention. Sale (1970), on the basis of observations on the distributions of larval Acanthuridae off Hawaii, also suggests that eddies contribute to larval retention. In describing the differences in distributions between larval fish species and planktonic crustaceans in relation to oceanographic processes around Hawaii, Leis (1982) proposes diverse mechanisms for both active larval retention for certain species and passive for others (at least on the spatial scale of his observations). He concludes that oceanic current eddies by themselves are unlikely to be a sufficient mechanism for retention in that they are not stationary. He also concludes that for the species exhibiting an inshore-neritic distributional pattern, "relatively active swimming by the animals" is required in conjunction with "a tidal eddy and possible nearshore upwelling." Again, the recurrent theme that may be generalized is the existence of specific behavior patterns at diverse life-history stages in relation to physical oceanographic processes, patterns that effectively retain the pelagic phases in an appropriate geographic area to ensure population persistence.

Sale (1980, pp. 372–379) reviews the literature on the life-history characteristics of coral reef fishes. He concludes that it is not yet possible to generalize on the spatial scales of populations and the associated spatial scales of larval retention and drift. Nevertheless, for some species there is evidence that active behavior is associated with physical oceanographic processes rather than passive drift.

Larval studies within the coastal zone also indicate that active behavior is involved in maintaining appropriate distributions. Marliave (1981) makes a connection between the vertical migration of larvae of the soft sculpin, *Gilbertidia sigalutes,* and retention. Recent studies of larval stages of near-shore fish off southern California by Barnett et al. (1984) describe behavioral differences as being related to diverse physical oceanographic processes that enhance retention. Marliave (1986) demonstrates limited planktonic dispersal for several fish species inhabiting the rocky intertidal zone in the northeastern Pacific Ocean. He argues that behavioral mechanisms minimize offshore losses and, possibly, longshore dispersal.

The timing of events in the life history has been linked to retention. Lobel (1978) suggests that the spawning of some Hawaiian fish is timed to correspond to periods of high levels of mesoscale turbulence in order to increase the probability that the larvae will return to their spawning site (mesoscale includes tens to hundreds of kilometers). Johannes (1978) and Watson and Leis (1974) report that fish reproductive activity around Pacific islands is timed to correspond to periods of low net current velocities. Thus, the temporal pattern in reproductive behavior has been interpreted to enhance larval retention. Barnett et al. (1984) suggest that the timing of the planktonic stage of the fish *Seriphus politus* is associated with the season of maximal cross-shelf water motion caused by internal tides. They infer that such motion in relation to ontogenetic changes in behavior enhances larval retention in the appropriate geographic area for population persistence.

There is also a concomitant loss of larvae that are diffused and advected away from the appropriate distributional area for the population. However, Thresher and Brothers (1985) observed no obvious relationship between length of the larval phase for Pacific oceanic island marine species and breadth of distribution:

> Analysis of the relationships between duration of the pelagic larval stage (as indicated by otolith microstructure), adult size, and extent of geographic distribution for Indo-West Pacific angelfishes (Pomacanthidae) indicates that, contrary to widespread assumptions, neither adult size nor larval duration sig-

nificantly correlates with distribution, either individually or jointly in a multiple regression.

They also suggest, as do Strathmann (1982) and Palmer and Strathmann (1981) for benthic invertebrates, that in pomacanthids "selection for long distance dispersal is unlikely to be the principal factor underlying the evolution of a long duration pelagic larval stage."

This lack of evidence for colonizing ability due to duration of larval phase indirectly supports the argument that it is retention that is being selected for in the duration of the larval phase rather than dispersal for colonization. The longer larval phases for a particular oceanic island species may be a function of the extent of the physical system used for retention (i.e., the larvae must travel farther to get back to the natal island) or poor growth conditions within the larval retention area (i.e., the larvae pass slowly through this phase of the life cycle). Victor (1986) suggests that the ability of a coral reef fish to delay metamorphosis may be an adaptation for maximizing the return of a planktonic larvae to coastal waters.

The above summaries of population persistence in what could be considered special environments address the first aspect of the member/vagrant hypothesis, the processes regulating the very existence of populations of species having complex life histories in the oceans (as well as the particular population pattern and richness). For both estuaries and widely separated oceanic islands we conclude that certain aspects of life histories (timing of spawning, larval behavior, ontogenetic changes in behavior, duration of larval phase) are "spatial" adaptations to permit population persistence in a particular geographic space.

There is evidence from another special marine environment, the coral reef, that may support the third aspect of the hypothesis, i.e., the regulation of the temporal variability in population abundance. The coral reef fish literature is not in complete agreement about the importance of competition for energy resource in the regulation of population abundance. Findley and Findley (1985) conclude that populations of butterfly fish of different species are not resource limited. Williams (1980) finds no evidence for resource limitation of pomacentrid reef fish populations. He concludes that abundance is a function of events during the planktonic phase of the life history and that the population question should be looked at in relation to the life cycle as a whole.

Although Talbot et al. (1978) consider coral reef fish at the community rather than the population level, their remarks are nevertheless pertinent to generalizations on population regulation of reef fish. Competitive interactions between species, they state (p. 425), appeared unimportant in explaining the distribution of species between reefs. They suggest that "nonequilibrium conditions are characteristic of coral reef fish communities and that because of these conditions, high within-habitat diversities are maintained." Robertson et al. (1981), in their experimental study on the availability of space for the territorial damselfish (*Eupomacentrus planifrons*), conclude that abundance is below the carrying capacity of the environment: "The results of both experiments indicated that a surplus of suitable substrate existed in the experimental areas and that *E. planifrons* was not generally limited by the availability of habitable space."

Victor (1983), from an analysis of recruitment patterns of the bluehead wrasse (*Thalassoma bifasciatum*) on coral reefs in the San Blas Archipelago (Caribbean coast of Panama), concludes, "[B]ecause recruitment did not reflect patterns of mortality on the reef, the findings do not support the view that reef fish populations are resource-limited." Doherty (1983), using an experimental approach, has also concluded that population abundance of territorial damselfishes (*Pomacentrus wardi* and *P. flavicauda*) are not resource limited. His experiments "support a nonequilibrial view of reef fish communities, rather than hypotheses that emphasize the importance of competition among populations usually at carrying capacity" (p. 176).

Although the literature on benthic populations in the open ocean is not extensive, there is some evidence of egg and larval retention for commercially important benthic species on continental shelves: the studies by Lough (1976) for *Cancer magister*, Nichols et al. (1982) for *Cancer pagurus*, Johnson (1960) for *Panulirus interruptus*, Johnson et al. (1984) for *Callinectes sapidus*, Efford (1970) for *Emerita analoga*, Makarov (1969) for several decapods, and Sinclair et al. (1985b) for three scallop species. Nichols and co-workers, however, interpret their results in relation to drift, not to retention.

The question of spatial scale is important to the difference in interpretation. As has been the case for zooplankton (see Chapter 7), the definition of the spatial integrity of populations has not been considered an important question for marine benthic species. Recent studies by Davidson et al. (1985), Campbell and Mohn (1983), and

Sinclair et al. (1985b) suggest that discrete populations of (respectively) snow crab, American lobster, and several species of scallop do exist. Geographically discrete aggregations of sea scallops, which may be self-sustaining, are illustrated in Figure 6.2. Coherence in landings of American lobster within several geographic areas during this century suggests that there may be several self-sustaining populations in the Atlantic provinces of Canada (Figure 6.3). The underlying mechanisms controlling the inferred population definition have not been described.

In sum, where the physical discontinuities in the ocean can be readily visualized (such as in estuaries and oceanic islands or reefs), the evidence is substantive—particularly in estuaries—that pelagic phases of the life cycle of marine species can be retained in a particular ocean space to permit persistence of populations. In these special marine environments population pattern and richness are clearly a direct function of spatial constraints. In addition, there is evidence that abundance, for some reef fishes at least, is not resource limited. The coral reef fish literature on population regulation suggests that energetics processes are not critical, but that spatial processes, as defined in relation to the member/vagrant hypothesis, may be important.

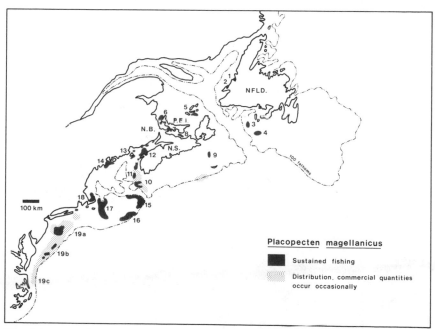

Fig. 6.2   Distribution of *Placopecten magellanicus* fishing areas.

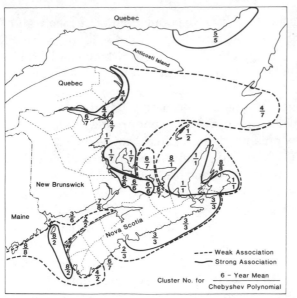

Fig. 6.3 Areas sustaining geographic populations of American lobster. The areas are associated statistically on the basis of coherence in the temporal trends in landings over several decades. (Reproduced from Campbell and Mohn 1983)

...7...

OCEANIC ZOOPLANKTON POPULATIONS

The literature on even the existence of populations of zooplankton in the oceans is scant compared with what has been written about other biota, such as marine fish and terrestrial animals. It was noted in Chapter 2 that during the late 1800s and early 1900s there was considerable interest in intraspecific geographic population patterns, particularly by naturalists in Germany, which eventually led to the new systematics involving the *Rassenkreis* and the biological species concept. However, the population richness of marine zooplankton was not extensively studied at that time. Other than the studies by Schmaus (1917) and Schmaus and Lehnhofer (1927) on *Rhincalanus cornutus*, it was not until the 1950s that detailed studies on the population structure of marine plankton were carried out (see the introduction of Fleminger and Hulsemann 1973). Fleminger (1973), in the context of evaluating the difficulties in the systematics of *Eucalanus*, makes the point that "population thinking" itself is poorly developed in marine plankton studies. There is, he says, a "widespread lack of an adequate basis for viewing the planktonic taxon in the perspective of a biological population."

Geographic Distribution

The zooplankton species for which at least some aspects of population structure have been described are listed in Table 7.1. Two points are made. First, self-sustaining populations of zooplankton species do exist, and have in several cases been described on the basis of both life-history distributional data and morphological differences. Second, in many cases the discontinuities between populations parallel geographic discontinuities.

Brinton (1962) identifies the several distinct morphological forms of the euphausiid *Stylocheiron affine* in the Pacific Ocean as geographic races (or, following the terminology of this paper, as popu-

Table 7.1 Zooplankton studies in which population richness has been described or inferred.

Study	Species
Allen 1959	*Pandalus borealis* (shrimp)
Berkes 1976	Several euphausiid species
Brinton 1962	Several euphausiid species
David 1955	*Sagitta gazellae* (chaetognath)
Fleminger 1967	*Labidocera jollae* (copepod)
Fleminger 1973	*Eucalanus subtenuis* (copepod)
Fleminger and Hulsemann 1974	*Pontellina plumata* (copepod)
Fleminger and Hulsemann 1977	*Calanus helgolandicus* (copepod)
Hulsemann 1985	*Drepanopus forcipatus* (copepod)
Jones 1965	*Candacia pachydactyla* (copepod)
Jones 1969	*Thysanoessa longicaudata* (euphausiid)
Karnella and Gibbs 1977	*Lobianchia dofleini* (small pelagic fish)
Kulka et al. 1982	Several euphausiid species
Lee 1971, 1972	*Centropages typicus* (copepod)
	Temora longicornis (copepod)
Lindley 1980	*Thysanoessa inermis* (euphausiid)
	Thysanoessa raschii (euphausiid)
Mauchline 1960	*Meganyctiphanes norvegica* (euphausiid)
Mauchline 1965, 1966	*Thysanoessa raschii* (euphausiid)
McGowan 1963	*Limacina helicina* (pteropod)
Schmaus 1917; Schmaus and Lehnhofer 1927	*Rhincalanus cornutus* (copepod)

lations). At least two of the forms, the "California Current" form and the "central" form, are associated with current systems and identifiable water masses. The euphausiid *Thysanoessa longipes* is found in the Pacific Ocean as two forms: the "spined" form is more restricted to sub-Arctic waters, being predominant in the Gulf of Alaska, while the "unspined" form is distributed within the North Pacific Drift and the California Current (Brinton 1962, p. 201). Geographically distinct populations of *Euphausia gibboides* (Brinton 1962) have been described for the North Pacific Drift and for the Peru Current (even though, morphologically, individuals from the two populations are similar) (Figure 7.1). Similar geographic discontinuities have been described by Brinton in the distributions of putative populations of *Euphausia eximia* (Figure 7.1), *Euphausia recurva*, *Euphausia brevis*, and *Thysanoessa gregaria*. David (1955) has described morphologically different populations of the chaetognath *Sagitta gazellae* on different sides of a frontal

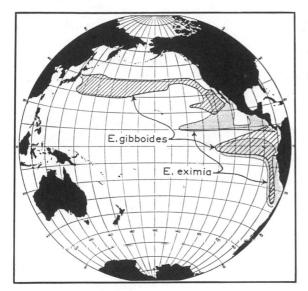

Fig. 7.1 Distribution of populations of *Euphausia gibboides* and *E. eximia* in the Pacific. (Reproduced from Brinton 1962)

zone between current systems in the sub-Arctic Ocean. Fleminger (1973) infers from morphological and distributional data that there are discrete populations of *Eucalanus subtenuis* in the Atlantic, Pacific, and Indian oceans. He states, "Grouped samples falling within the hydrographic limits of any one ocean do not differ significantly . . . and serve to emphasize the stepped or discontinuous pattern of the means from the different oceans."

From the distributional data of Fleminger and Hulsemann (1977 and references therein) it can be inferred that there may be several self-sustaining populations of *Calanus helgolandicus* in the Mediterranean Sea, the Black Sea, the Middle Atlantic Bight, the North Atlantic Drift, and the North Africa upwelling zone. The same authors have also described separate populations of *Pontellina plumata* in the Atlantic and Pacific/Indian oceans (Fleminger and Hulsemann 1974). Jones (1965) provides morphological evidence that there are separate populations of the copepod *Candacia pachydactyla* to the east and west of southern Africa. Hulsemann (1985) has identified separate populations of *Drepanopus forcipatus* in the Southern Ocean on the basis of morphometric and distributional information. Karnella and Gibbs (1977), also on the basis of distributional and morphological data, infer the existence of discrete populations of the lanternfish (*Lobianchia dofleini*) in the eastern North Atlantic Central Gyre, the western North Atlantic Central Gyre, and the Mediterranean (the small size of this fish, maximum of 38 mm standard length, justifies including it in a discussion of zooplankton evidence). There may well be additional distinct populations of this species in the southern Atlantic, southern Indian, and eastern South Pacific oceans (Karnella and Gibbs 1977, p. 376).

McGowan (1963) describes two morphologically distinct populations of the pteropod *Limacina helicina* in the North Pacific Drift and in the North Pacific Sub-Arctic Gyre and notes a behavioral difference between the two populations: the population in the non-recirculating system (at least in the surface waters) migrates vertically, but the gyre-based population does not. This difference in behavior can be plausibly related to the constraint of population persistence in the two differing physical systems.

The above studies indicate the existence of more than one self-sustaining zooplankton population within the distributional limits of the species. In addition, Fleminger and Hulsemann (1974) have described several panmictic species in the genus *Pontellina*. They argue on the basis of morphological information that for several species there is but a single self-sustaining population within the distributional limits of the species, as is the case for European eel. They state (p. 111), "Absence of conspicuous geographical variation indicates sufficient transport and advection to maintain *panmixis* [emphasis added] within each species except the Atlantic and Indian-Pacific populations of *plumata*." They also discuss the physical oceanographic mechanisms that may sustain the particular geographical patterns of the various observed panmictic populations of the different *Pontellina* species (p. 112): "The circulation systems and physical conditions known to maintain these lenses of eutrophic tropical water and the pools of oligotrophic tropical-subtropical waters are the obvious mechanisms sustaining the geographical distribution of the four species of *Pontellina*." In three of the four cases the species distribution is also the population distribution.

In sum, populations of diverse zooplankton species apparently sustain themselves in relatively fixed geographic locations in the open ocean. In most of the above cases the mechanisms that achieve persistence in the face of diffusion and advection have not been well established; the emphasis, instead, has been on describing large-scale patterns from a biogeographical perspective. There is, however, a growing literature, generally on a much smaller spatial scale, dealing with the connection between zooplankton population persistence and physical oceanographic processes.

Population Persistence

Before reviewing the evidence on the mechanisms of zooplankton population retention, let us first consider some aspects of the pattern of distribution of coastal-zone copepod species. Fleminger (1967, 1975) and Fleminger and Tan (1966), in a rich series of publications, have described the distribution of coastal-zone planktonic copepods of the genus *Labidocera* in the western hemisphere. In Fleminger's 1967 report, the discontinuities between three of the species (*Labidocera jollae*, *L. diandra*, and *L. kolpos*) are defined in detail (Figure 7.2). The distributional discontinuities are sharp, with no sample yielding more than one species. In this coastal zone characterized by considerable advection and mixing (Inman and Brush 1973),

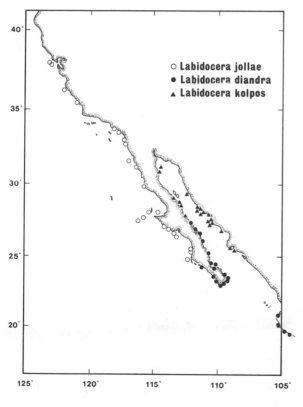

Fig. 7.2 Distribution of three species of *Labidocera* off the coast of California and Mexico. (Redrawn from Fleminger 1967)

○ **Labidocera jollae**
● **Labidocera diandra**
▲ **Labidocera kolpos**

geographically persistent patterns in zooplankton species distribution are sustained.

The detailed biogeographical studies by Fleminger on *Labidocera* are perhaps the most complete ever made of any zooplankton genus. The results indicate considerable biological structure, in this case at the species level, in a highly diffusive and advective environment not characterized by sharp physical boundaries. The *Labidocera* species must have life-history behavioral features that ensure retention within their appropriate coastal zone. Within the limits of distribution of the various *Labidocera* species there may well be a number of self-sustaining populations.

There is some evidence in support of zooplankton population retention in relation to physical oceanographic processes from the ecological (rather than systematics and biogeographic) literature on marine zooplankton. In studies dealing with the mechanisms of persistence of plankton species in particular geographic areas it is often inferred (but not always specifically stated) that a population is being studied. The discreteness of the group in morphological terms is not usually evaluated. The key ecological questions relating to the persistence (and retention) of zooplankton populations were somewhat paradoxically but clearly stated in 1905 by Damas, the ardent proponent of fish larval drift. By that time, it had already been observed that every plankton species has its own characteristic geographical distribution. Given the prevailing view of physical oceanography at the time, with its strong emphasis on large-scale circulation and mean flow, Damas asked: Why are the species of plankton animals in their distribution limited to distinct areas of the sea? Why is it that these animals, which we know may be carried far away with the currents, are not within a short time evenly distributed from pole to pole? Damas (1905, p. 22) concluded,

> The species maintains itself due to the existence in these areas of a circulatory current which periodically brings back a certain proportion of the individuals which are entrained in the continual movement of the water to the surface of the ocean. . . . The mechanism of the circulation plays, thus, the principal role in the conservation of the species in this area.[4]

Helland-Hansen and Nansen (1909, pp. 312–316) provided some support for Damas's conclusion by describing the recirculation of water and the water masses in the Norwegian Sea (the particular study area of Damas's zooplankton work). Somme (1933, 1934) further elaborated the above concepts in the same geographic area as well as in the Lofoten area of coastal Norway, but he was perhaps overly influenced by the prevalent physical oceanographic concepts and emphasized drift and dispersal of plankton away from the spawning areas. He did not address how temporal persistence is maintained in the face of such dispersal. The same emphasis on drift in relation to residual circulation was made by Redfield (1939) and Redfield and Beale (1940) in their distributional studies on the pteropod *Limacina retroversa* and several chaetognath species in the Gulf of Maine.

None of the above landmark studies attached much importance to vertical differences in the circulation to account for the persistence of the life-history distributions; the emphasis was on depth averaged surface-layer residual circulation. If a gyre was described, persistence in zooplankton distributions could be explained. If not, plankton were assumed to drift with the inferred residual current.

The contemporary reviews by Rose (1925) and Russell (1927) show that vertical migration of zooplankton species was actively studied throughout this early period. None of these studies, however, referred to the effect of vertical shear in the circulation on the horizontal distribution of a vertically migrating plankton, nor were the ontogenetic changes in vertical distribution and behavior linked to vertical circulation effects. The emphasis in these earlier studies on vertical migration of zooplankton was physiological (do animals regulate their migration in relation to light, temperature, salinity?) rather than distributional.

Vertical shear in the circulation in relation to persistence of distributions of zooplankton populations was first considered by Ottestad (1932) for copepods and Mackintosh (1937) for macroplankton in the Antarctic Ocean. Ontogenetic changes in vertical distributions of various life-history stage were hypothesized to ensure persistence of the populations (or species). Marshall (1979, p. 430) states, in a review of this literature, "In the Southern Ocean such ontogenetic migrations between deep winter and shallow summer levels tend to maintain populations within their proper bounds."

Later studies in which persistence of zooplankton distributions in particular geographic space has been "explained" by links between animal movement (ontogenetic and daily) and both vertical and horizontal physical oceanographic processes are summarized in Table 7.2.

Table 7.2 Zooplankton studies in which the mechanisms of population maintenance in restricted geographic areas have been described.

Study	Species	Location
Aldredge et al. 1984	*Calanus pacificus* (copepod)	California upwelling zone
Allen 1959, 1966	*Pandalus borealis* (shrimp)	Northumberland coastal waters, U.K.
Binet and Suisse de Sainte Claire 1975	*Calanus carinatus* (copepod)	North Africa upwelling zone
Boucher 1982	Several copepod species	North Africa upwelling zone
Boucher 1984	Several copepod species	Ligurian front
Boucher et al. 1987	Several copepod species	Ligurian front
Damas and Koefoed 1907	Several copepod species	Greenland Sea
Davis 1984a	*Pseudocalanus* spp. (copepod)	Georges Bank
Fleminger 1985	*Calanus pacificus* (copepod)	California upwelling zone
Kling 1976	Several radiolarian species	Eastern North Pacific
Kulka et al. 1982	Several euphausiid species	Bay of Fundy
Mackintosh 1937	Several macroplankton species	Southern Ocean
Mauchline 1960	*Meganyctiphanes norvegica* (euphausiid)	Clyde Sea
Mauchline 1985	Several euphausiid species	Rockall Trough
Ottestad 1932	Several copepod species	Weddell Sea
Peterson et al. 1979	Several copepod species	Oregon upwelling zone
Petit and Courties 1976	*Calanus carinatus* (copepod)	North Africa upwelling zone
Rothlisberg and Miller 1983	*Pandalus jordani* (shrimp)	Oregon upwelling zone
Wroblewski 1982	*Calanus marshallae* (copepod)	Oregon upwelling zone

Particularly noteworthy are the studies by Peterson et al. (1979) and Wroblewski (1982), which describe in considerable detail the mechanisms of retention of zooplankton populations from several species within the upwelling zone off the Oregon coast. Prior to these studies it was presumed that zooplankton abundances increase either downstream or offshore from the main upwelling center due to drift in relation to the residual circulation. However, Peterson et al. (1979, p. 468) conclude,

> Through various behavioural responses, each of these dominant copepods [*Calanus marshallae, Pseudocalanus* sp., *Acartia clausi, Acartia longiremis,* and *Oithona similis*] has a different mechanism by which its population is maintained in the upwelling zone. Some populations complete their entire life cycle within 10 km of the shore, where offshore transport is minimal. Others develop further offshore but return to lay their eggs close to shore. For all copepods, reproduction and naupliar development occur in locations that minimize seaward advective losses.

As they summarize in their abstract, "The population of each zooplankton species appears to be maintained within the upwelling zone by a specific relationship between its distribution and the circulation." Peterson et al. (1979) propose that diel vertical migration at particular life-history stages of *Calanus marshallae* reduces transport out of the upwelling zone. Wroblewski (1982), in a simulation model, reproduces most of the empirical observations on *C. marshallae* by Peterson et al. (1979). The simulations demonstrate how diel vertical migration by adults can interact with upwelling currents to increase residence time of egg-laying copepods in the near-shore zone. For the other copepod species described in this study, however, diel migration was not observed, even though ontogenetic changes in vertical distribution were taken to be important in maintaining persistence within the upwelling zone.

Aldredge et al. (1984), in a field study using a submersible, observed a persistent aggregation of Stage V copepodites of *Calanus pacificus californicus* at a depth of 450 m, about 100 m above the bottom of the Santa Barbara Basin. Density of these juvenile copepods was estimated to be as high as 26 million per cubic meter. The animals were in a state of diapause. Aldredge and co-workers relate the diapause stage in deep water to population persistence in dynamic, seasonally variable upwelling systems. Citing the life cycle of another species, they state:

> Diapause at depth by *C. carinatus* has also been associated with annual upwelling cycles off the coast of Africa. Early developmental stages of this species inhabit upwelled surface waters of the Guinea Current and adjacent waters. At the end of the upwelling season, Stage V copepodites descend to below 500 m

and remain there until the onset of the next upwelling. Current patterns at the depth of diapause have been found to flow toward or parallel to the coast and thereby act to conserve the resting Stage V copepodites within the region of seasonal diatom blooms.

The life cycle of these two copepod species (*C. pacificus californicus* and *C. carinatus*) is thus a function of constraints imposed by the physical geography.

Boucher (1982) describes persistent distributions of several populations of copepod species in two upwelling areas off Cap Ghir (Morocco) and Cap Blanc (North Mauritania). From the comparative results of two sampling strategies (fixed station observation versus sampling along the path of a drogue) vertical migration is shown to be important in maintaining the observed spatial distributions for several species (*Centropages, Candacia,* and *Pleuromamma* populations). It is further inferred that populations of *Euterpina, Corycaeus,* and *Temora* also use vertical migration behavior to maintain a persistent distribution in spite of the residual circulation. Boucher concludes (p. 205) that the study "provides evidence that behavior resists dispersion of the organisms by the currents," and (p. 206) that the "results show the degree of relative independence of zooplankton to entrainment by a current."

Boucher (1984) in a similar study describes the spatial pattern of zooplankton populations in relation to the Ligurian haline front off Nice, France. Again, persistent geographic patterns for populations of *Calanus helgolandicus, Euchinella rostrata, Pleuromamma gracilis,* and *Cavolinia inflexa* are explained by interactions between species-specific behavior and dynamic features of the environment. Differences in distributional responses to the same environmental features lead Boucher to conclude that differential behavior is important in generating the observed persistent spatial patterns (ontogenetic migration for *C. helgolandicus* and *E. rostrata,* diurnal migration for *P. gracilus* and *C. inflexa*). He states (p. 482), "Maintenance of a fixed localization by the cohorts despite the complex dynamics of the water masses and the geographical displacements of the front itself implies active swimming by the organisms."

The follow-up study by Boucher et al. (1987) in the same area considers a broader group of zooplankton taxa and evaluates the changes in the physical structure over an annual cycle, as well as the impact of the changes on the zooplankton spatial distributions. Similar conclusions are drawn. Species-specific behavior in relation to particular aspects of the physical environment generates diverse persistent spatial patterns on an annual time scale. They state:

> This second set of data confirms that copepods with similar morphology (*Centropages typicus* and *Temora stylifera*, for instance) do not respond in the same manner to the same dynamical environment. For these reasons hydrodynamical features are insufficient to explain the observed distributions. Other ecological requirements and behavior are involved.

Kulka et al. (1982), in a study of several species of euphausiids (including *Meganyctiphanes norvegica*, *Thysanoessa inermis*, and *Thysanoessa longicaudata*) in the Bay of Fundy, note that the vertical migratory behavior (ontogenetic and diel) of populations of each species is different in different geographic areas. They conclude (p. 333), "The vertical migratory pattern of each population is different and is adapted to maintain position in separate local hydrographic conditions." Further, they suggest specifically that vertical migration patterns are mechanisms to ensure persistence of populations in particular geographic space.

It is tempting to generalize that the complex patterns observed in ontogenetic and diel vertical migrations of marine zooplankton species are partially mechanisms for population persistence in particular geographic space. Hardy and Gunther (1935) and Hardy (1953; 1967, chap. 15) suggest essentially the opposite, that zooplankton migration in relation to vertical current shear is an adaptation for dispersal although vertical migration may also enhance persistence in particular areas. They recognize the role of vertical migration in population persistence but do not stress it. The vertical migration behavior has been interpreted to allow avoidance of poor food environments and thus efficient sampling of a patchy food resource (Isaacs et al. 1974). Evans (1978) has elaborated on Hardy's hypothesis using a simulation approach. Other generalizations have stressed the role of predator avoidance (Zaret and Suffern 1976; Ohman et al. 1983; Gliwicz 1986) or of optimizing metabolic processes in relation to thermal stratification (Mclaren 1963; Enright 1977; Enright and Honegger 1977). Koslow and Ota (1981), however, in a study of the vertical migration of *Calanus pacificus*, *Rhinocalanus nasutus*, and *Euphausia pacifica*, conclude

(p. 108), "No single hypothesis can explain the diverse migratory behaviour of these zooplankters." Longhurst (1976), in a relatively recent review of vertical migration, draws the same general conclusion.

However, a "population persistence" hypothesis of vertical migration is not inconsistent with the remarkable complexity of behavioral patterns both between populations of the same species (e.g., McGowan 1963 and Kulka et al. 1982) and between species (as reviewed by Longhurst 1976). One would expect diversity in behavior patterns within and between species in light of the wide range in oceanographic environments within which the many zooplankton populations are distributed. Such a hypothesis cannot be rigorously evaluated at this time, but it is consistent with many of the empirical observations cited in this essay.

The part played by behavior, including vertical migration, in generating aggregations associated with reproduction is an important one. Mauchline and Fisher (1969), in their exhaustive review of the biology of euphausiids, discuss the problems of maintenance of populations in particular geographic areas and the critical importance of behavior. They state that for *Meganyctiphanes norvegica* (p. 347), "Aggregations for breeding with subsequent summer dispersions of the populations appear to be a feature of the distribution of this species within any one geographical area; it may also be a feature of many more species but, at present, adequate data are lacking." In their opinion, swarming, at both surface and subsurface (which is a component of vertical migratory behavior), must play an important role in maintenance of population structure (p. 350):

> As Sir Alister Hardy suggested many years ago, the possibility that social behavior of these animals plays a considerable role in maintaining populations cannot be excluded, not only in coastal areas but in the open ocean. The general area of occurrence is apparently defined by the water masses but many species occur in two water masses and so scope for *social mechanisms of population maintenance* is obviously present. (Emphasis added)

The reluctance of biological oceanographers to accept that planktonic species have complex behavior, coupled with the heavy emphasis on passive drift, is reflected in a short note by Rice and Kristensen (1982) on surface swarms of crab megalopae. They describe an aggregation of megalopae several kilometers across, with the animals "restricted to the near-surface layer but distributed fairly evenly in a horizontal plane at densities of perhaps 100/m². " They stress (p. 238):

> Of much greater interest than the origin of the megalopae is the fact that they had maintained or re-established dense aggregations after such long periods of pelagic life. . . .
>
> It is difficult to imagine what possible advantage there could be in the portunine megalopae forming dense aggregations either during the pelagic phase or at the time of settlement which would certainly have occurred within a few days. It seems much more likely that the aggregations were *chance happenings.* (Emphasis added)

It is stretching credibility to interpret as a product of chance the cohesiveness suggested by these observations, which imply a well-organized behavioral activity of the animals. From the perspective of the life-cycle continuity necessary for persistence of populations within particular geographical constraints, it is not difficult to imagine possible advantages to swarming. Williams and Conway (1984), in a detailed study of vertical distribution of *Calanus helgolandicus* in the Celtic Sea, state that the importance of behavior has been underrated (p. 63): "The seasonal and vertical migrations of *C. helgolandicus* are part of a more complex pattern of inherent behavior than has been reported previously . . . however difficult this is to discern in the natural populations, it always expresses itself."

It appears that the inherent behavioral pattern varies with maturation stages. Harris (1963) has described by experiment the differences in vertical migration behavior of *Calanus finmarchicus* in summer and winter. During the summer experiments, under conditions of constant light or constant dark the copepods continued to migrate vertically. Animals collected in the winter months, however, remained at the bottom of the experimental containers and did not demonstrate vertical migratory behavior.

Foxton (1964), in an analysis of seasonal variation of zooplankton distributions in Antarctic waters, supports the earlier conclusions of Ottestad (1932) and Mackintosh (1937) concerning vertical migration: "It may be as Mackintosh (1937) has suggested that the most important result of migration is of *navigational importance* in that species utilize the northerly flowing surface water by summer and the southerly flowing warm deep current by winter in such a way that the bulk of the population is maintained within its optimum geographic range." Van der Spoel and Heyman (1983), from a biogeographic perspective, also conclude that vertical migration is important in population maintenance in particular geographic space (p. 167):

> Vertical migration is executed by many representatives and this migration is brought forward in this book as a behavior warranting the *geographic stability of populations*. Consequently, when explaining phenomena in plankton, one should always consider both the ability, and the inability of a specimen to maintain a stable geographic position. . . . Geographic locality, hydrological conditions and specific behavior may lead to completely different results of shift in different populations. In some cases the vertical migration may add to the horizontal stability, in other cases it does just the reverse: stimulate the horizontal displacement of populations. (Emphasis added)

Marshall (1979, p. 431), with reference to deep ocean plankton, addresses the issue of population persistence which was so clearly defined earlier by Damas:

> The life-history strategy must sustain the populations of a species in the face of a moving environment. Clearly, the pattern must enable enough of the young to grow up in the right ecological surroundings. Current speeds are likely to be greater at mesopelagic than at bathypelagic levels, but many species from both zones spend their early life in the still livelier surface waters. How do species keep in their living space?

Marshall subsequently summarizes the literature on the diverse life cycles in deep-ocean pelagic species that permit population maintenance in particular geographical space in the face of diffusion and advection. Many of the life cycles involve daily, seasonal, or ontogenetic vertical migration.

There is thus a small but persistent literature, beginning with Damas and Koefoed (1907), that has considered the importance of vertical migration to population persistence within the constraints of a diffusive pelagic environment. The diverse extant hypotheses, however, have tended to attribute this ubiquitous behavioral pattern to energetics processes (primarily in relation to food consumption and predator avoidance). We point out here that the geographical patterns in the distribution of morphologically defined zooplankton populations have frequently been associated with physical oceanographic systems

(Table 7.1). Also, several detailed studies have concluded that ontogenetic changes in behavior are associated with population persistence in particular geographic space in relation to constraints of the local physical oceanographic processes (Table 7.2). Finally, we suggest that vertical migration behavior in marine zooplankton may frequently be associated with population persistence, rather than (or in addition to) being an adaption to energetics processes (food, predation).

Relative Abundance

A final component of the zooplankton literature is pertinent to the general hypothesis under consideration. It concerns the control either of relative abundance of individual zooplankton populations or of the biomass of particular size fractions of the zooplankton as an aggregate.

Several recent studies have concluded that continental shelf zooplankton populations are not food limited. Mclaren (1978) finds that growth dynamics of seven species of copepods from Loch Striven, Scotland, are not food limited but can be described by temperature alone. Davis (1984b) concludes that the phytoplankton standing crop on Georges Bank is unlikely to limit growth and reproduction in *Pseudocalanus* sp. or *Paracalanus parvus*. Ohman (1985) draws the same conclusions for a *Pseudocalanus* sp. population in Dabob Bay, Washington. Longhurst et al. (1984) observed that in the eastern Canadian archipelago many Arctic zooplankton species descend to deep water prior to the slowing of algal production in the near-surface waters, which suggests that copepods are not constrained by food availability.

However, several studies have demonstrated food limitation of biological processes but not of population abundance itself (Landry 1978; Huntley 1981; Runge 1981; Frost et al. 1983; Peterson 1985). Huntley and Boyd (1984), in a synthesis of the zooplankton literature, conclude that the growth of marine particle-feeding zooplankton is not likely to be limited by the availability of food in the coastal regions. They infer that zooplankton abundances in the open ocean are likely to be food limited.

Evidence from systematics studies, however, suggests that some oceanic copepod species are not food limited. Frost and Fleminger (1968), in their revision of the genus *Clausocalanus*, discuss the lack of differences in feeding appendages between sympatric species (p. 89):

Also noteworthy is the absence of interspecific differences in the structure and armature, though not in the size, of the feeding appendages. Despite significant differences in the shape of the forehead and rostrum, the oral appendages were found to be highly conservative and uniform throughout the genus. There is morphological evidence that the species of *Clausocalanus* are all particulate grazers feeding in a similar way. . . . This evidence also suggests that the sympatric species of *Clausocalanus* are not feeding competitively on a limited food supply.

The overall literature is difficult to interpret in relation to population regulation. Generally the evidence for food limitation involves growth of individuals and fecundity, not absolute abundance of the population itself. The benthic literature provides examples in which the control of individual growth rates is decoupled from the control of numbers (Ebert 1968, 1982; Spight 1974). In this sense, demonstrating that growth and fecundity are food limited, for example, does not necessarily mean that abundance itself is food limited. However, the contrary observation—that growth and fecundity are not food limited—does logically lead to the conclusion that abundance cannot be food limited.

Thus, in these cases at least, where food limitation of growth and fecundity processes has been convincingly ruled out, population numbers must be controlled by other factors. Predation is usually put forward as the most probable alternative. Ohman (1985) states this as follows (p. 15): "The food limitation paradigm is not applicable to *Pseudocalanus* sp. for which predation is a more likely mechanism of population regulation." Both Ohman (1985) and Davis (1984c) argue that predation is sufficient to regulate the observed population levels of herbivorous copepods (in Dabob Bay and on Georges Bank, respectively), but they do not evaluate the role of advective and diffusive losses. An alternate mechanism of regulation that usually is not considered involves spatial processes, i.e., losses from the *appropriate* geographic area for the population in question. There is support for this mechanism of population regulation of marine zooplankton from the Continuous Plankton Recorder program.

The interannual variability in zooplankton abundance in the northeastern Atlantic has been analyzed in considerable detail by Colebrook (1978, 1979, 1981, 1982a–c, 1984, 1985) and Colebrook and Taylor (1984). Colebrook (1982a) has described the annual fluctuations

for the years 1958–1980 of various plankton species in the northeastern Atlantic and the North Sea. Persistence is observed in the geographic distributions of diverse species even though there are considerable seasonal and interannual differences in relative abundance and a significant downward quasi-linear trend. Colebrook (1985), in an examination of seasonal variations in the pattern of year-to-year fluctuations in abundance of *Pseudocalanus elongatus, Acartia clausi, Calanus finmarchicus,* and hyperiid amphipods in the North Sea, concludes that population abundances are defined during the winter and are related to "inherent limitations in rates of population increase and possibly to generation times." Over the thirty-five years of observation there has been a coherent gradual (15%–50%) decline in numbers of essentially all populations of zooplankton species in the northeastern Atlantic and the North Sea (Colebrook 1985, table 1). The analysis suggests that "there may be a relatively weak link between average levels of primary and secondary production."

The accumulated observations suggest that both the trend in zooplankton abundance (which is coherent over a very large spatial scale) and the interannual variability (which is controlled by events during the winter when many of the species are not feeding) may be controlled by advective losses from the appropriate distributional area during the period of reduced growth and diapause. Control by predation is possible, but one would not expect to observe coherence between essentially all species in their downward trend due to predator-prey interactions, nor to observe that abundance is controlled during the winter season for all species. However, one would expect coherence between species if population control is primarily a function of the physics of retention of populations in particular geographic space. Also, if advective losses from populations of zooplankton are determining the trend and the variance, the impact of such loss could reasonably be expected to be most critical during the winter season of limited or no growth in numbers.

In sum, it has been concluded on the basis of the accumulated continental shelf zooplankton literature that food is frequently not limiting. There is some evidence from the only large space/time scale study on zooplankton population dynamics (the Continuous Plankton Recorder program in the northeastern Atlantic) that physical oceanographic processes are critically important in the control of abundance.

Concluding Remarks

In what sense do the diverse studies of marine zooplankton support the member/vagrant hypothesis? Populations of zooplankton species have been identified in only a few cases. In most cases the geographic extent of the populations is associated with well-defined physical oceanographic systems. Also, in a few studies persistence in particular geographic space has been related to both ontogenetic and daily changes in vertical distribution. Studies have involved persistence of populations on large scales such as the Southern Ocean, as well as on smaller scales such as the Oregon and California upwelling areas. There is a small literature, but one of long duration, that has considered the behavior of plankton (particularly vertical migration) as important to population persistence in the face of the physical constraints of diffusion and advection.

Population pattern and richness, to the degree that they have been described in the open ocean, seem to be readily explained by spatial processes. This is not a new generalization, and is perhaps to be expected. However, when joined to the fisheries literature the generalization is of interest. It strengthens the conclusion that population pattern and richness for fish can be defined at the planktonic egg and larval stages in relation to the same sorts of physical oceanographic constraints that generate pattern in zooplankton populations. In addition, given the greater knowledge of population richness in marine fish and the association of their early life-history stages with, in some cases, small oceanographic features (see Chapter 5), it is probable that populations of zooplankton exist on smaller spatial scales than generally have been studied. The geographical extent of zooplankton populations has not been a question of interest to marine ecologists, and as a result it is poorly described and understood.

A different body of the zooplankton literature approaches regulation of abundance indirectly. Usually neither the geographic scale of the population is known, nor is abundance itself described. Rather, density (numbers per unit volume) is studied in relation to food availability. It has been generalized that for coastal and continental shelf zooplankton food is usually not limiting for growth and reproduction. The lack of evidence for food limitation of such growth processes strongly suggests that other factors must be involved in the regulation of abundance. The empirical observations resulting from the most substantive study of interannual variability of zooplankton abundances in both shelf and open-ocean environments in the northern Atlantic have been interpreted in relation to physical oceanographic processes (*without* food-chain linkages). These studies are consistent with the member/vagrant hypothesis. Other workers, however, have favored predator control of abundance when food limitation cannot be demonstrated. Undoubtedly, the direct roles of physics on vagrancy and predators on mortality are important in the regulation of abundance. It is perhaps premature to compare their relative importance.

···8···

SUMMING UP THE EVIDENCE

In summarizing the evidence to support the member/vagrant hypothesis, we first make two points:

- The literature just cited in Chapters 5 to 7 as evidence for the member/vagrant hypothesis of population regulation has not been arranged in an appropriate form to test null hypotheses. The questions addressed, the hypothesis defined, and the evidence cited constitute in many respects a consistency argument about an alternate view of population regulation in the oceans.

- The hypothesis does not negate the importance of food limitation or predator control of numbers in marine populations where they are shown to be critical. Rather, it infers that such factors (which are defined here as energetics processes) need not act in a density-dependent manner for population stability. The hypothesis does infer, however, that for marine species having complex life histories, spatial processes predominate in the regulation of the four aspects of populations that are addressed.

Four characteristics of marine populations have been discussed: richness, pattern, absolute abundance, and temporal variability. In support of the hypothesis being introduced, several subject areas have been considered: fisheries populations, estuarine populations, oceanic island and reef fish populations, and zooplankton populations.

In no single subject area is the literature sufficiently comprehensive that the evidence in support of the general hypothesis is convincing. For example, zooplankton population richness and pattern have been poorly studied, but there are several detailed studies on the mechanisms by which zooplankton populations sustain themselves in particular geographic areas by interactions with physical oceanographic processes. In contrast, the literature on population richness and pattern for marine fish is well developed, but the bulk of the

literature (see Harden Jones 1968; Cushing 1975) stresses the importance of dispersal at the egg and larval phases by drift with the residual
currents and thus does not support the hypothesis. We infer that the
deeply engrained emphasis on fish egg and larval drift has been self-
fulfilling to the degree that the sampling designs have frequently been
defined to demonstrate drift and the empirical observations have
sometimes been misinterpreted.

The literature as a whole (zooplankton to fish, estuaries to the
open ocean), however, is mutually supportive. In geographic areas
where the obvious constraint on population persistence is the ability
not to be advected or dispersed away from the oceanic island or the
estuary, the specialty literature is full of examples of how populations of
the various species do indeed manage to persist.

The fisheries literature clearly demonstrates that there are
marked differences between species in population richness. There is a
continuum in this characteristic of species from Atlantic salmon, for
example (highly population rich), to Atlantic cod (moderately population rich), to Atlantic mackerel (population poor), to European eel (a
single population throughout the distributional limits of the species). It
is clear that population richness is defined at the planktonic phase of
the life history rather than at the juvenile or nonreproductive periods of
the adult phases.

The geographic discontinuities at the early life-history stages,
in conjunction with the ability of adults to home to particular
oceanographic features, define population structure. The evidence is
not so complete as one would want for any particular species, but when
a comparative approach is taken the evidence is persuasive. For Atlantic salmon, American shad, and striped bass, species whose egg and
larval phases are completed in a particular river system, that river
system itself supports a self-sustaining population. For smelt, whose
eggs and larvae drift from the individual spawning rivers to a common
downstream estuarine or coastal zone larval distributional area, the
complex of rivers flowing into the larval retention area (rather than the
individual rivers) is the geographic unit which supports a self-sustaining population.

The mirror images of Atlantic salmon and European eel life
histories forcefully make the point that it is the early life-history
geographic discontinuity/continuity features that define population
richness rather than nonreproductive adult geographic discontinuity/

continuity features. Adult eel distributions are markedly discontinuous prior to their spawning migration, yet the species is population poor; adult Atlantic salmon distributions are continuous during parts of the oceanic phase of the life cycle, yet the species is population rich. In the past the anadromous/catadromous life-history distributions have been considered, not as part of a life-history continuum, but as a paradox (Baker 1978). From the continuum perspective introduced here the underlying constraints on population richness (in this case from one to many populations) are simply interpreted. Population richness, which the evolutionary synthesis interprets as speciation in action, has not previously been accounted for in the ecological literature.

Perhaps it is useful to view the population richness characteristic of marine species having a planktonic stage from a plankton perspective. The other life forms of a population (postlarval, juvenile, and adult phases) are, from this view, parts of a cycle that are necessary for the long-term temporal persistence of the larval distribution in particular geographic space. This shift in emphasis from adult to larva puts homing, which permits population richness, into a different perspective. Homing of adults to specific spawning sites allows temporal persistence of the larval distribution in particular geographic space. The weak link in the cycle of many species may be the planktonic phase. The ability of the egg and larval components of the life cycle of a population to persist and maintain an aggregated distribution at shorter time scales (weeks to months) is the key, in this view, to the population richness question. The spatial scale on which the planktonic phase of the life cycle can be coherent, and the corresponding number of physical oceanographic (or geographic) systems underlying that coherence within the distributional limits of the species, defines the degree of population richness. Further, it is the particular behavioral characteristics of the species (both ontogenetic changes and those within a particular stage of the life cycle) in relation to the physical oceanographic features of the environment that define the spatial scales of coherence in the planktonic phase of the life cycle. It is clear from comparing the closely related species of Atlantic cod and haddock (cod being considerably more population rich) that subtle differences in behavior must be involved.

It is this area—mechanisms of persistence of plankton distributions in particular geographic space (be it an upwelling zone, the fringes of an oceanic island, or an estuary)—that is well developed in the flanking literature to fisheries biology. Representative studies from the estuarine (Table 6.1), the oceanic island marine fauna, and marine zooplankton (Table 7.2) literature amply demonstrate that plankton can persist in fixed geographic locations in spite of the average residual circulation. It is inferred that the animals are much more influenced by the variability in the circulation and mixing characteristics of their particular environment, and the vertical structure of these features, than they are by the depth-averaged residuals. Platt and Harrison (1985), in a very different context (resolving the order of magnitude differences in estimates of open-ocean primary production by two different methods), stress the "crucial importance of giving at least as much emphasis to the variances as to the means in discussing non-linearly coupled systems."

In sum, the fisheries literature, in particular, tells us that population richness is defined at the planktonic phase of the life cycle. The zooplankton literature on population ecology and the special environments literature on estuaries and oceanic islands (which are amenable to detailed study) tell us how plankton can use behavior in conjunction with the particular physics of the area of distribution to persist in a highly diffusive environment. Life-history features (such as timing of spawning, duration of the larval stage, and the existence of diapause) are interpreted to be associated with physical oceanographic constraints to population persistence in particular geographic space. From consideration of the combined literature, both questions of the regulation of population richness and pattern can be answered in general terms. Control of absolute abundance and its temporal variability can perhaps be considered as corollaries of the hypothesis accounting for richness and pattern.

The existence of self-sustaining populations, and the geographical pattern of the population structure of a given species, is, we argue, a function of the ability of the planktonic phase of the life cycle to maintan an aggregated distribution in the face of dispersal by diffusion. The spatial scale of the dispersal/retention balance and the specificity of appropriate spawning areas for particular populations, then, may partially define absolute abundance of the population. The characteristic mean abundance of a population, following this argument, may be physically constrained.

Much of the evidence to support this part of the general hypothesis is indirect. Mean abundances of populations of Atlantic herring,

Atlantic cod, and haddock in a qualitative sense are a function of the size of the larval retention area, rather than a function of features of the distributional areas of the juvenile and adult phases (which are frequently shared among populations of different levels of abundance). Further, marine fish larvae have infrequently been shown to be food limited. Rarely, however, can the *losses* from the appropriate distributional area be separated out from mortality due to predation. As such, the relative importance of spatial processes versus energetics processes in generating the total losses from a population cannot be evaluated on the basis of the present literature. Further, continental shelf zooplankton appear in many cases not to be food limited. For the most comprehensive study on interannual variability in zooplankton abundance, the predominant influence is argued to be large-scale variability in ocean circulation and mixing.

For the northeastern Atlantic, persistence in particular geographic areas of overwintering components of populations (both continental shelf and open ocean) is critical to both the seasonal dynamics of the populations and the long-term trends. Analysis of this massive overall data set on zooplankton abundance has led Colebrook (1985) to conclude that there are weak links between phytoplankton and zooplankton dynamics. If continental shelf zooplankton are in many cases not food limited, other processes must be involved. Again, predator control may be sufficient, as has been argued by Davis (1984c) for Georges Bank zooplankton populations and by Ohman (1985) for Dabob Bay. An alternate controlling factor of absolute abundance and its variance is the direct control by spatial processes.

Brander and Dickson (1984) have observed that Atlantic cod density (abundance per square kilometer at fishable sizes, i.e., juveniles and adults) is considerably higher in the North Sea than in the Irish Sea. Yet, cod growth rates between the two areas are very similar. The egg and larval distributional area for cod in the Irish Sea is characterized by a restricted seasonal envelope of phytoplankton growth as well as by markedly higher diffusion caused by geographic differences in the dissipation of tidal energy. Given the above arguments that fish larvae are not food limited, it may well be that geographic difference in the physical oceanography of the spawning and early life-history distributional area accounts directly for the differences in absolute abundance between seas.

Absolute abundances of populations with complex life histories are interpreted here to be defined at sexual reproduction in relation to spatial constraints for continued membership of the progeny in the population. If the hypothesis is robust, vagrancy is at times density dependent. Following this line of argument and the literature cited earlier, we argue that absolute abundance is largely physically constrained. Variability around the inferred physically constrained mean level of abundance has been repeatedly observed in diverse fish and invertebrate populations to be associated with variability in advection and, presumably, mixing.

As already argued for marine fish, the deeply engrained assumption that the role of the larval phase in species with complex life histories is for passive dispersal has influenced the manner in which benthic ecologists have studied populations and their regulation of abundance. The pelagic phase of the life cycle has very often been considered as a "black box," and the source of recruitment to the study site as an unknown. The spatial scale of much of the work being done in benthic population studies may be inappropriate for the questions being addressed in this essay. If the spatial scale of the population is in reality hundreds of kilometers of shoreline, a single study site in any particular location within the distributional area may not encompass the appropriate scale of environmental parameters (either biotic or abiotic) that are in fact regulating absolute abundance and temporal variability. Caffey (1985), in a study of spatial and temporal variability of intertidal barnacles, concludes that variable recruitment from the larval phase is critical to the population dynamics of benthic species with complex life histories. The benthic ecologists may frequently be working on too small a spatial scale. At the other extreme, the population geneticists and paleobiologists may be working at too large a spatial scale.

The assumption that larvae of benthic species are widely dispersed is being re-evaluated by population geneticists (see Hedgecock 1986 for an overview) and is at present a point of contention. In this essay, we argue that the genetic structure of marine invertebrate populations should not be inferred from apparent dispersal capabilities of the life history. Because of these questions of scale in benthic invertebrate studies, which are partially a function of the assumption of passive drift at the larval stage, the whole question of geographic self-sustaining populations has received short shrift. Nevertheless, there are studies in the benthos literature that support the hypothesis (cited in Chapter 6).

Complex life histories are predominant in the marine environment. Given the initial constraint of sexual reproduction, life histories at the population level must be geographically constrained (membership requires being in the appropriate geographic location). Complex life histories are usually described in relation to morphology, and the life history can be presented as a morphological cycle ("history" unfortunately emphasizes the linear time element rather than the spatial cycle). In the perspective being pursued here, the cycle concept is stressed; the complex life history has both a morphological cycle and a geographical cycle. The two cycles, through behavior, permit population persistence. Simply put, complex life histories are the mechanism by which populations persist in particular geographic space. The complex life histories (or cycles) generate the persistent patterns induced by the constraint of sexual reproduction. In that these constraints are considered to be fundamental, the population regulation hypothesis developed from them (even though developed predominantly on the basis of evidence from the pelagic environment) should be of general significance in the oceans.

There is an overall trend within the evolution of sexually reproducing animal species toward the collapse of complex life histories. Garstang (1929) describes this temporal trend for the molluscan phylum in some detail. He states (p. 80):

> When we pass from the more primitive and ancient groups of Mollusca to the more modern ones, the larva no longer hatches as a simple trochosphere, but is provided with a shell and foot from the first, and the simple girdle of cilia which constituted the prototroch is replaced by a much more powerful organ, the *velum*. . . .
>
> There are of course many Gastropods and Bivalves in which, even under marine conditions, the free-swimming larval stage has been secondarily reduced in association with a marsupial or incubatory mode of development. . . . Finally, the pelagic stage may be suppressed altogether, and the whelk emerges from the confinement of its brood-chamber as a diminutive adult, ready at once to pursue its definitive career.

Garstang does not elaborate on the underlying causes of this trend through time other than to state (p. 81), "Although, with fuller knowledge of the facts and of the bionomical conditions, it may be possible to explain the cases of reduction or obliteration of the larval stage in terms of adaptation, it seems more probable that there has been a secular change tending to depreciate the value of dispersal as the sea became stocked with an increasing number and variety of specialised habitats."

We shall consider in Chapter 11 an alternate interpretation of the causes for the secular trend in the collapse of complex life histories. The requirement for life cycle closure within a changing spatial structure (and an emphasis on retention rather than dispersal) is considered important.

The point to be made here is that the relative importance of physical oceanographic processes and food-chain events in the regulation of abundance might be expected to vary as a function of the degree of collapse in the life history. Populations of species with free-living egg and larval stages may be predominately regulated by spatial constraints to life cycle closure. In contrast, food-chain constraints, under the member/vagrant hypothesis, would be predicted to be more critical for populations of species with collapsed life histories.

With the more complete internalization of the complex life history, as in sharks and marine mammals, for example, the constraint of a diffusive environment at one link in the cycle is removed or at least minimized. Population regulation (richness, pattern, absolute abundance, and variability) for these species categories may well be fundamentally different. It will be argued in later chapters, however, that some aspects of the hypothesis are of general significance in ecology and evolution, in both marine and terrestrial environments. Before addressing such implications it is useful to indicate a paradoxical aspect of the member/vagrant hypothesis, particularly in view of the key role of dispersion in terrestrial ecology.

There is a disturbing quality of the emphasis in this essay on retention in the hypothesis on population regulation. Some aspects of life cycles would appear to be adapted for dispersal, such as pelagic eggs; yet it is being argued that advection and diffusion are ultimate constraints in the definition of populations, as well as in the regulation of abundance. It may well be that these coupled processes cannot be considered separately, that they are two components of a single phenomenon. Holton (1978) has discussed the frequent occurrence of coupled phenomena in physics such as continuity/discontinuity. Such phenomena have to be considered together to be adequately understood. Here, the emphasis has been focused on retention rather than dispersal.

···9···

IMPLICATIONS FOR ECOLOGICAL QUESTIONS

The member/vagrant hypothesis addresses four aspects of population regulation: two species-level characteristics (pattern and richness), and two population-level characteristics (mean absolute abundance and temporal variability). To understand why a species separates into self-sustaining populations in particular spatial patterns is important when one addresses the questions of abundance and temporal variability for a single population. The constraint of sexual reproduction is central to pattern and richness.

Life cycles may be considered spatial mechanisms that ensure persistence given the particular constraints of the physical geography. The life cycle has both morphological characteristics and particular spatial or geographic settings. This conception of life cycles may be considered to be an ecological extension of Bonner's (1965) treatment in his essay, *Size and Cycle*. He states (p. 3):

> The view taken here is that the life cycle is the central unit in biology. The notion of the organism is used in this sense, rather than that of an individual at a moment in time, such as the adult at maturity. Evolution then becomes the alteration of life cycles through time, genetics the inheritance mechanisms between cycles, and development all the changes in structure that take place during one life cycle.

In Bonner's synthesis, genetics, evolution, and development are tightly linked in his focus on life cycles, but the ecological framework is left out. Yet, many features of complex life cycles are interpretable only in relation to the constraints of the physical geography. Closing the cycle to ensure sexual reproduction in an appropriate geographical setting for subsequent continuing closures is in a certain sense a struggle against the diffusive and advective processes in the marine environment. The constraint of sexual reproduction and the concomitant requirement for life-cycle closure in particular geographic space are, in the member/vagrant hypothesis, the driving forces that generate pattern and richness.

From this perspective, before studying the factors that are important in the regulation of abundance of a particular population of a sexually reproducing species, it is essential to first identify the spatial integrity of the population. Is in fact a population being studied at all in sampling a particular study area? Individuals of several different populations of the same species can be mixed together during parts of the life cycle (this is a problem for mobile animals). It is critical to define the appropriate spatial scale of the study in relation to the geographical distribution of the population. For capelin, say, which home to a broad area of coastline extending over hundreds of kilometers, the early life-history events in any single bay may not reflect the distributional area of the population as a whole.

The scale of investigation has to match the scale of the phenomena being addressed. If the geographical extent of the population is not known, this needs to be recognized and the implications thought out. Regulation of abundance of the *population* cannot be understood without first understanding its geographical extent. The importance of various processes affecting losses from the population may, however, be addressed at a range of spatial scales below the scale of the population distribution itself. The debate over the relative importance of density dependence versus density independence almost always involves investigations of processes on these smaller spatial scales. This method of investigation may in itself have contributed to much of the confusion in the literature.

Energetics and Spatial Processes

In the member/vagrant hypothesis two kinds of processes involved in the regulation of abundance of individual populations are defined: energetics processes and spatial processes. Both can involve density-dependent and density-independent population events. However, if there is density dependence in the spatial processes for a given population (for example, if the vagrancy rate increases with increasing abundance), abundance can be controlled without evidence of density dependence in energetics processes.

There is a tendency to associate all energetics processes with density dependence without verification. This tendency is misleading. Pepin (1985) and Kean (1982), in experimental studies of predation on fish larvae, did not observe density-dependent mortality, and Taggart

(1986) did not observe density-dependent mortality in a detailed field study on capelin larvae. Clearly there can be density-independent mortality due to starvation, predation, and disease. The opposite phenomena, density-dependent spatial losses from a population (such as reproduction in the wrong area for the population at higher levels of abundance), may be less well recognized in the marine literature.

The total accumulated *loss* of individuals in a given year class of a population from the egg stage to maturation and successful sexual reproduction comprises (1) spatial loss from the population (which does not imply immediate mortality), (2) death due to starvation as a result of competition for food resources, and (3) death by predation and disease. For the oceans it is inferred, from the indirect evidence cited in the previous chapters, that spatial loss (which involves physical processes and behavior) frequently predominates in the regulation of population abundance in the oceans. Competition for food resources or avoiding predators may generate a relatively small component of the overall loss empirically observed from egg to successful sexual reproduction. Fecundity, in this sense, is considered to be predominantly a function of accumulated life-cycle losses due to spatial processes rather than predominantly a function of intraspecific competition for limited resources or food-chain events.

In succeeding chapters the argument is developed that the inferred predominant mode of population regulation for marine species with complex life histories (spatial losses from the population at various stages of the life cycle) is critical to the process of *speciation*, whereas the mortality generated by competitive phenomena is critical to ener-getics-related *adaptation*. Speciation and adaptation in this treatment may be decoupled even though both phenomena involve aspects of population regulation. If the member/vagrant hypothesis is valid, in-traspecific competition for limited resources can play a minor role in the population biology of many marine species, in particular those with complex life histories. MacCall (1984) has rigorously developed the dual role of spatial and energetics constraints on the population biology of northern anchovy in the California Current. Many of the points that are elaborated in descriptive terms here have been worked out in a quantitative sense in MacCall's study.

Much of competition theory has come from first developing theoretical equations of population growth (based to a large degree on data from human societies) and then testing the theoretical relationships by experiment, frequently using nonsexually reproducing species (the first two experiments testing the equations were made on the bacteria *Escherichia coli* and the yeast *Saccharomyces cerevisiae;* see Hutchinson 1978, p. 23). Following the simple development of concepts introduced here, competition for limited resources would be logically invoked for populations of species with the asexual mode of reproduction. The bifurcation that occurred with the evolution of sex (i.e., the constraint of relation between individuals of the same species within the time frame of a generation) radically changed the "numbers game" and imposed new geographical constraints as well: where you are suddenly became perhaps more important than how competitive you are.

The development of competition theory in ecology by first observing human populations and then testing on asexual organisms may well have been based on false premises about the generality of the observations. In addition, many of the classic experiments designed to test the role of competition for limited food resources in population regulation—whether they used asexual species (bacteria and yeast) or sexual species (*Drosophila melanogaster,* the water flea *Moina macroscopa, Paramecium* spp., the flour beetle *Tribolium confusum*)— involved bounded physical systems in which the hypothesized critical role of spatial loss of individuals from the population was eliminated by the experimental design (see Hutchinson 1978, pp. 21–28, for a historical review of this literature). In this respect the experiments may not be applicable to population regulation in the oceans. Since such developments were fully consistent with the putative role of intraspecific competition for resources in evolutionary theory, some comfort undoubtedly was derived from the overall coherence between the two disciplines.

In that ecological generalizations (in particular, generalizations about regulation of population abundance) are used generally without qualification in flanking disciplines such as population genetics, biogeography, paleontology, and developmental biology, it is important that the role of competition be satisfactorily resolved. It is more than an in-house quarrel. The wrong generalizations involving population regulation and competition may have pervaded the literature in the above-mentioned fields; one can cite, for example, Mani (1982) in population genetics, Valentine (1985) in paleobiology, and Brown and Gibson (1983, chap. 3) in biogeography. The language of modern ecology, particularly as expressed by Hutchinson, is seductive but may well be based on false premises.

Life-History Theory

The member/vagrant hypothesis has implications for some aspects of marine ecology other than population regulation. Life-history theory, which is a blending of population regulation and the life-tables of the demographer, emerged in the 1960s as a result of the seminal paper by Cole (1954). Within ecology it has become an identifiable field of study within which fitness is measured somewhat in the abstract (i.e., usually without reference to spatial or geographic constraints). The field has been hailed as a success story in modern ecology (Stearns 1982) because of its power in elucidating diverse life-history phenomena. The life-history traits frequently evaluated in relation to fitness are life span, age at maturity, fecundity at age, and age-specific mortality rates. With the use of optimization approaches and the implicit assumption that there is competition for limited resources, trade-offs between growth and reproduction in diverse life histories can be analyzed.

The hypothesis presented here suggests that there may be fundamental problems with the life-history theory approach, in the oceans at least. The evidence, as it has been marshalled, does not support the assumption that competition for limited resources is necessarily the major factor in the regulation of population abundance.

More importantly, what is left out of life-history theory may be critical to the understanding of specific traits. Just as population genetics has tended to ignore the phenotype, life-history theory to a large degree ignores geography. Complex life histories in this essay are considered mechanisms for population persistence within specific geographic constraints. In this perspective the morphological cycle from egg to adult is paralleled by a geographic space cycle within which population persistence is possible. The geographical constraints are viewed as being critical to such life-history features as life span, age at maturity, and timing of spawning. The distances to be traveled for life-cycle closure of the population, for example, within the geographical/morphological cycle may have important implications on whether semelparous or iteroparous reproduction, or whether a short or a long larval phase, is characteristic of a particular species.

It seems fruitful when interpreting life-history features to consider the simple developmental constraints of size—that, given fixed cell division, it takes longer to be bigger (Bonner 1965)—along with the geographical constraints of population persistence introduced in this essay. Life-history theory, to the degree that it may be based on false premises (in those cases in which competition for limited resources is not critical) could, because of the ingenuity of the practitioners, generate reasonable explanations for incomplete reasons when applied to marine species. Also, more complexity than is required may be invoked in the interpretation of empirical observations of life histories.

Two additional areas of marine ecology to be discussed here in relation to the population regulation hypothesis involve perceptions in biological oceanography. There is a strong emphasis in biological oceanography on analyzing communities in relation to fluxes of energy and chemical constituents. Population dynamics, which involves looking at numbers of individuals of particular species in addition to energy, is not well developed. Platt (1985) discusses the utility of the Sheldon biomass spectrum (Sheldon et al. 1972, 1973, 1977) in marine ecology, relative to other conventional approaches that involve the identification of organisms to populations of species. Platt appears to agree with Bahr (1982), who asserts that the premises of systematics "are obsolete, having been established before the theory of evolution: many existing species have yet to be named; at best its application in nature is subjective." Platt (1985, p. 62) adds:

> Two other weaknesses are of direct relevance here. One relates to a further aspect of its obsolescence: it is out of step with that school of ecological thought that attaches importance to the macroscopic view and recognizes that organisms must obey basic thermodynamic laws. The second is simply that a conventional taxonomic description of a typical pelagic sample is highly demanding on time. And not to put too fine a point on it, after the job is done the investigator is not always certain what to do with the result.

These conclusions by a respected marine ecologist perhaps reflect a particular trophic-dynamics school within biological oceanography. Are they well founded? It is not the utility of the Sheldon biomass spectra approach to marine ecology (which has more than proven its worth) that is taken issue with, but rather the imputed obsolescence of conventional population ecology which involves identifying living particles at some higher level than their diameter. The perspective on population regulation developed here implies that the conventional approach is far from obsolete if the aim of marine ecology includes understanding such fundamental questions as patterns in living particles in time and space. The patterns that are empirically observed are hypothesized to be generated by the constraint of sexual reproduction and the challenge of physical oceanographic processes to life-cycle closure of populations of species with particular behavior. Such patterns, particularly if food is not limiting, may not be based on energetics constraints. Consideration of populations of particular species may not be out of step with either the big picture or thermodynamic constraints. There are many marine ecological questions that may be most efficiently addressed by the Sheldon particle spectrum approach, but this does not imply that the conventional approaches of population dynamics involving the identification of individuals to species are obsolete. The energetics approach (including the analysis of particles) on its own may not be sufficient to account for features of the macroscopic view.

The second point addresses perceptions on the stability of spatial patterns in the oceans. The member/vagrant hypothesis infers persistence of populations of marine organisms in specific geographic space in relation to physical oceanographic processes on an ecological time scale up to decades. Steele and Henderson (1984), in contrast, conclude, "The assumption by fisheries management of a natural persistence in stocks [i.e., populations] is questionable." They base their conclusion on a limited selection of empirical observations on decadal variability in biological properties in the oceans (with a heavy emphasis on pelagic fish) and a model which includes variable physical forcing, density-dependent regulation at high abundance, and predator regulation at low abundance. Steele (1985) in a subsequent review article amplifies the above conclusion.

We take issue with the selection of initial empirical observations with which the model results are compared. Steele and Henderson (1984) and Steele (1985) chose highly selective examples from the literature to represent biological temporal variability in the oceans. They point out that Atlantic herring landings off the west coast of Sweden have been observed to vary dramatically in essentially an all-or-none sense (at pulses of a few decades) for several hundred years. However, relative constancy has also been observed in herring landings that track recruitment variability at dozens of other precise spawning sites in the northeastern Atlantic over much of the same period. The Swedish herring was, and still is, recognized as an anomaly in herring fisheries variability (as is the Plymouth herring; Cushing 1961). Does one emphasize the dozens of herring spawning populations that have shown persistence in their spawning location and relative biomass (interannual variability well within an order of magnitude until purse-seining became the dominant fishing method), or, as Steele and Henderson have done, the two populations which come and go?

A second example used by Steele and Henderson to illustrate biological instability is the so-called Russell Cycle observations off Plymouth at the mouth of the English Channel (Southward 1980). Plankton species composition, hydrographic variables, and fish species composition observed at this fixed station flipped in the 1930s and then in the 1970s returned to almost the pre-1930 condition. However, this geographic area is coincident with a biogeographic boundary. At tens of kilometers north and south of the fixed station, changes in at least the species composition of the commercial fish landings have not occurred (unfortunately, there are no comparable time series on plankton and hydrographic variables). The Russell Cycle may be an artifact of sampling very close to a biogeographic boundary which oscillates spatially to a certain degree. A fixed station off Nantes or Liverpool would have shown persistence in populations (at least for commercially important species) of constant species composition in the landings, albeit a different mix between the two locations.

That considerable variability is a characteristic of marine populations is not questioned, nor is the evidence that eastern boundary currents in particular are characterized by species composition changes on an ecological time scale; but we do wish to balance the record. Huxley (1881) argues that herring spawning populations have persisted off the Yarmouth area for centuries (p. 612):

> There is historical evidence that, long before the time of Henry the First, Yarmouth was frequented by herring fishers. . . .
>
> As long as records of history give us information herrings appear to have abounded on the east coast of the British Islands, and there is nothing to show, so far as I am aware, that, taking an average of years, they were ever either more or less numerous than they are at present.

The fisheries in question were based on spawning populations that home to very precise locations. In other words, in this area at least, population persistence in particular geographic space has been well documented. Huxley also makes the point, as do Steele and Henderson, that in other areas (in particular off Bohuslän) this temporal persistence is not observed. Mackerel spawning has occurred annually in the northwestern Atlantic in two well-identified continental shelf areas since before the 1900s. Separate persistent haddock populations have been observed scientifically since the 1920s on Georges Bank and Browns Bank. Populations of Atlantic cod have been observed on the Grand Banks for hundreds of years, and *Calanus finmarchicus* populations have been spawning regularly in the Lofoten since plankton nets were designed. Colebrook has observed persistence in plankton populations in the northeastern Atlantic since the 1940s. Even in highly dynamic eastern boundary currents, population persistence may occur in specific geographic locations. Devries and Pearcy (1982) state that, off Peru, "the distribution of anchoveta and hake vertebrae in surface sediments roughly parallels the distribution of modern day populations." In addition, they observe biogeographical shifts in spatial boundaries rather than internal shifts in species composition during glaciation/deglaciation changes.

These examples of persistence of population patterns in highly dynamic physical environments need equal treatment in the Steele and Henderson analysis. If the disruptive effects of fishing are taken into account and the anomalies are considered as anomalies, a strong case can be made for persistence of marine populations in time and space. To the degree that the biological conclusions of these authors are dependent on the examples selected, they are in our view suspect and not supported by much of the literature cited in this essay.

Terrestrial Environments

To what extent are the concepts developed in the population regulation hypothesis applicable to the terrestrial environment? As indicated above, evolution of higher forms has repetitively involved internalization of complex life cycles. To a certain degree, it can be generalized that evolution on land for vertebrates has in this way insulated populations from many of the geographical constraints that exist for complex life histories in the oceans. Nevertheless, for many terrestrial species, including primitive man, geographical population structures are well described. Populations in many, if not most, species are associated with specific geographic space, particularly during reproduction. As in the oceans, population richness varies between species (as summarized by both Mayr and Rensch). Some of the processes regulating populations (especially the generation of population pattern and richness) may be common to both marine and terrestrial environments.

During the late nineteenth century and first few decades of the twentieth, systematists devoted much more attention to terrestrial populations (especially birds) and their geographic patterns than to the marine species. The key difference between the two systems, terrestrial and marine, has been the subsequent developments within ecology of the study of these populations. Fisheries ecologists, due to the management implications of the population concept, have since the work of Heincke, Hjort, and Schmidt incorporated the populations of the systematists into their ongoing research and have gone beyond the simple morphological description of populations which dominated population systematics.

This has not occurred in general terms in terrestrial ecology. The existence of population richness (and the differences between species in this characteristic) and some aspects of the geographical patterns were well described. However, with a few notable exceptions, little is known about the persistence in time and space of the populations, their absolute abundance, or their temporal fluctuations. Population dynamics studies for terrestrial species have monitored population density (numbers of individuals per unit area or volume) more frequently than absolute abundance; except for terrestrial ecologists who manage wild game, they simply do not have an analog of the port sampling scheme of the fisheries biologist which allows the populations to be studied at the appropriate time and space scales of the biological phenomena.

In sum, there has not been in ecology, for terrestrial environments, the logical outgrowth of the paradigm shift of new systematics, i.e., population systematics. To a certain degree terrestrial ecologists still implicitly use the typological species concept of the early nineteenth century in their day-to-day activities (Kingsland 1985). Individual variability is recognized, but population thinking in Mayr's terms is not the norm.

Given the lack of appropriate evidence for the persistence of terrestrial populations and their absolute abundance and temporal variability, it is difficult to determine the degree to which the general population hypothesis for the oceans is applicable to land. The empirical observations on animal movement do imply that geographical constraints of populations of many terrestrial species parallel the ocean situation. It has been repeatedly documented that dispersal distances (i.e., the distance between birthplace and breeding place) are skewed toward short distances, a surprising fact given the mobility of the species studied. Most individuals move a short distance from their birthplace; only a few move farther. Murray (1967), in a simple model involving space limitation for mating and behavior, reproduced the skewed dispersal distance observations for vertebrates.

A sampling of almost every issue of an ecological journal contains examples supporting limited dispersal of individuals of terrestrial species. Trexler (1984), in a study of a chrysidid wasp species, observed that only one of several hundred marked *Trichrysis tridens* migrated between undisturbed host population sites. Newton and Marquiss (1983) in their study of the dispersal of sparrowhawks between natal and breeding areas, report that most of the birds settled to breed in the general area where they were born, and recoveries fell off rapidly with increasing distance from the birthplace. They also found that changes in the median dispersal distance between years were not correlated with sparrowhawk population densities, suggesting that competition does not influence distribution.

There is indirect evidence that dispersal out of the appropriate distributional area is important in the regulation of populations in terrestrial environments. Emlen (1986), for example, reports on a comparative field test of the hypothesis that bird communities are resource limited (p. 125): "A fundamental concept of population biology, generally taken for granted in current ecological research, holds that the consuming biomass of animals inhabiting a delineated area tends to stabilize at or fluctuate around a level determined by the availability of resources, notably living space and food." He evaluates this concept by comparing what he calls "matched habitat patches" within continents and on islands. He observes that bird biomasses are higher on islands than on similar areas on continents, and thus infers that resources cannot be limiting. The bounded nature of islands appears to generate a fence effect which minimizes dispersive losses from the populations. The argument is similar to that developed in this essay for populations in the oceans. Vagrancy of birds on islands from their populations is minimized by well-marked physical discontinuities (the encircling water/land boundary).

The member/vagrant hypothesis, because of its roots in the richness and pattern characteristics, may help clarify the repeated observations on limited dispersal for terrestrial species. Baker (1978) has produced an exhaustive synthesis of animal movement in all environments. In it he develops the familiar-area hypothesis to account for the observation on return migration (p. 853):

> All terrestrial vertebrates and some invertebrates . . . seem to spend the major part of their lives within a familiar area, whether or not they perform a seasonal or ontogenetic return migration. . . .
>
> Paradoxically . . . the best evidence in support of the familiar-area hypothesis is found for birds. Avian migration is so spectacular, takes place over such long distances, and in many cases is so linear, that at first sight it might seem unlikely that it can be accommodated within a model based on the familiar-area thesis.

A major difficulty that Baker has in generalizing his hypothesis is the evidence from the oceans. Supposed dispersal at the egg and larval phases of marine fish, for example, is inconsistent with his hypothesis. Baker does not view the phenomenon of animal movement from a population dynamics perspective. From such a perspective the present synthesis, which emphasizes egg and larval retention at appropriate spatial scales for population persistence, may complement Baker's thesis.

Shields (1982, 1983) does look at animal movement from a population dynamics perspective, albeit with an evolutionary bent. He addresses the general question of why birth-site philopatry is so prevalent in the animal kingdom. A joint consideration of the general marine population hypothesis defined here and the rich presentation of both concepts and data by Baker and Shields leads to a simple interpretation of both return migration and philopatry in terrestrial environments. First, however, the essence of Shield's hypothesis on philopatry needs to be presented.

After reviewing, and finding unsatisfactory, other models of philopatry, Shields (1983) introduces his own genetic model. He suggests (p. 148) that "inbreeding can be an adaptive response to spatial heterogeneity in the *genetic* environment independent of ecological scale" (emphasis in original). In essence he argues that small differences in the genome "induce enough outbreeding depression to favour increased inbreeding." Thus, following concepts of natural selection that are now accepted, philopatry is selected for. Following these arguments, he states (p. 150), "Population subdivision and isolation by distance characteristic of so many organisms, in spite of their enormous capacities for dispersal and increased panmixia, finally becomes explicable."

In the genetic model of Shields, increased inbreeding itself is the primary function of philopatry, and increased outbreeding the function of "vagrancy." Shields (1983) tests his genetic hypothesis on philopatry/vagrancy by analyzing relative fecundities in philopatric and vagrant species. He states (p. 152): "Philopatry, which is effective in producing relatively intense inbreeding should occur in, and only in, low-fecundity organisms. Wider outbreeding resulting from vagrant dispersal should be limited to high-fecundity organisms."

The literature on the fecundity and life-history features of benthic marine invertebrates, as reviewed by Thorson (1950), supports the hypothesis. However, in essentially all of the studies that Thorson reviewed it was *assumed* that the pelagic egg and larval phases are adaptations for large-scale dispersal. The possibility that eggs and larvae are retained in a relatively restricted geographic area to ensure persistence of populations was not considered. In this sense, then, the benthic invertebrate literature is unfortunately not a rigorous test of the hypothesis. Shields (1983, p. 154) reflects current thinking when he states, "Normally, one can assume that pelagic larvae [of marine invertebrates] will disperse vagrantly as they are passively carried by local currents." He continues:

> Many of the marine fishes, like their invertebrate counterparts, are very fecund (10^6 and 10^9 offspring/lifetime). These high-fecundity species normally produce pelagic eggs, free-swimming larvae, or both. Their pelagic propagules are then subject to passive dispersal in water currents.

From the review of the marine literature in this essay, and the general population hypothesis generated, the above generalizations on philopatry in relation to fecundity do not adequately reflect the marine observations. Fecundity, following the member/vagrant hypothesis, is a function of the diffusive nature of the particular pelagic environment within which the population life cycle persists. In highly diffusive environments high fecundity is required to ensure population persistence. In less diffusive environments, or for marine fish species that attach their eggs to the bottom (such as herring), a lower fecundity will permit persistence; the fecundity of herring, for example, is lower than cod fecundity. It is argued here that most marine species, irrespective of their relative fecundity, are philopatric.

The observation that most sexually reproducing terrestrial animal species are characterized by both return migrations (Baker 1978) and philopatry (Shields 1982, 1983) is a function of fundamental ecological constraints. Persistence of populations of sexually reproducing animals demands philopatry, which is a manifestation of the biological species concept itself. The degree of inbreeding (and the general reduction of fecundity) may be a result of the collapsing of the complex life history and resultant shrinking of the geographical space scale within which populations persist, rather than a result of natural selection due to "outbreeding depression."

The genetic model of Shields and the familiar-area model of Baker are fully consistent with the life-cycle ecological constraints we suggest here. High fecundity, however, in this hypothesis, is primarily a function of geographical constraints (i.e., level of diffusion) at particular phases of complex life cycles, rather than a function of the degree of outbreeding being selected for (as hypothesized by Shields). The characteristic spatial scale of diverse populations, which is a function of the life-cycle geographical constraints, will produce a continuum in the degree of inbreeding. Very small populations with life cycles associated with small geographic space will be characterized by a higher level of inbreeding than larger populations (even of the same species) with life cycles associated with larger geographic space. Inbreeding itself, following these arguments, is not selected for, but is a result of the particular constraints of the spatial scale of the population's life cycle.

In short, some elements of the member/vagrant hypothesis of population regulation in the oceans may be relevant to terrestrial systems. Spatial patterns in terrestrial populations, and their absolute abundances, have not in general terms been accounted for in the terrestrial ecological literature, nor has the trend in the collapse of life cycles by the internalization of early life-history stages. The spatial fabric on which population persistence through life-cycle continuity is maintained on land clearly involves the structure generated by plants and trees as well as the physical geography itself.

This is a fundamental difference between oceanic and terrestrial environments. In the pelagic environment in particular, physical features predominate in generating population pattern, richness, absolute abundance, and variability. Other species, however, are still of considerable importance in permitting life-cycle continuity and population persistence. Red hake juveniles inhabit living scallops, and haddock larvae are frequently observed in association with jellyfish, and parasitism as a mode of existence by definition involves other species for population persistence.

Nevertheless, for the pelagic environment it is probable that physics (including the characteristics of the bottom) predominates over biological structure as the spatial fabric for population regulation in the broad sense used in this essay. In the terrestrial environment spatial structure generated by living material is no doubt of critical importance in population regulation. It is emphasized here that the spatial dimension itself, both physical and biological, in relation to the constraint of free-crossing of individuals within populations, may be critical to terrestrial population regulation. For asexual species, however, energy would be expected to be the factor of ultimate importance in the regulation of abundance.

Van Valen (1973a, 1973b, 1976) offers a rich and imaginative view of ecology and evolution in which competition for food is a core component. The basic arguments in support of the ultimate control of abundance by food availability are presented in his discussion of pattern and the balance of nature (Van Valen 1973b). The conclusions from his analysis are, however, an important component of the other two papers, whose concepts include "The Red Queen's Hypothesis," which has generated considerable discussion in the ecological and evolutionary literature. He argues (1973b, p. 36) that two aspects of distributional pattern strongly suggest that terrestrial herbivores are food limited (his analysis assumes that there is little disagreement at other trophic levels that abundance is food limited):

> Some ubiquitous patterns apply to terrestrial herbivores as well as to other animals and strongly suggest that the ultimate regulation of this trophic level as a whole, like that of other trophic levels, is by resources, in this case food. The patterns are that there are more small animal individuals than large ones, and that communities with greater primary productivity have more animals.

He is referring to density (numbers of individuals per unit area or volume) in the above generalization on numbers. He cites other evidence in support of food limitation, but these two "patterns" are the key components of the case. He concludes that it is nevertheless disturbing that evidence for food limitation for terrestrial herbivores is intuitively very hard to accept; in short, "the world is green," whereas if food is limiting, one expects to see some denuded forests. He states (p. 41): "The problem is sufficiently acute that I give it a name, the Enigma of Balance: How can it be that some species regulated even ultimately by food do not periodically greatly reduce their food by overeating?"

There is, however, no enigma if food is not usually either ultimately or proximally limiting to population abundance of sexually reproducing species. It is the evidence of food limitation (the two patterns identified above) that may have been misleading. First, the existence of more small animals than large animals does not necessarily imply or demand food limitation. Rather, it may be a product of spatial scale and the constraint of necessary encounter for sexual reproduction. Sexually reproducing small animals must be more numerous (i.e., have higher density) if sufficient encounter is to occur to ensure reproduction and population persistence. In this sense numbers in relation to size are physically constrained. Sexual reproduction generates a cost to rarity that is size dependent. Food limitation is not the only logical explanation of this empirically observed pattern. Deductive logic can be dangerous if key elements of the phenomena are left out (in this case space and relational constraints between individuals).

Second, is it indeed correct in the oceans that communities of higher primary productivity have more animals (not higher biomass of herbivores but more individuals in the component populations)? The population abundance of a copepod species in the oligotrophic North Pacific Gyre may be greater than the population abundance of a similar species on the eutrophic Georges Bank. Neither population is necessarily food limited. In the marine environment it is not well established that population abundances are higher in eutrophic than in oligotrophic waters. Even with respect to biomass there is some reason to conclude a weak coupling between trophic levels. Sinclair et al. (1984), for example, observed essentially identical trawlable fish biomasses in the Gulf of St. Lawrence and the Scotian Shelf, even though the gulf is considered to be more productive at the primary level.

More important, and equally relevant to the oceans and the terrestrial systems, is that higher primary production may well lead to higher biomasses at other trophic levels, but not necessarily to more abundant populations of the component species. Increased food availability in the system may lead to increased numbers of populations (and thus to a higher biomass than in less productive areas) rather than increased size of the individual populations. This subtle point is important in the context of the development of Van Valen's other papers on evolutionary biology. Population regulation can well be due to spatial phenomena and still be consistent with the observation that animal biomass as a whole is a partial function of photosynthetic energy. Higher levels of plant food in this sense may lead to more populations of different species (i.e., a higher diversity of populations and/or species) without the individual population abundances themselves being food limited. Speciation and adaptation are responsive to events at the population regulation level of organization rather than to overall energy availability.

In sum, deductions that are logical at the energetics level are not necessarily valid at the numerical level. Biomass at trophic levels may be ultimately food or energy limited. Does this imply at the population level that abundance is food limited, even ultimately? Van Valen argues in the affirmative. We say that the jump may not be valid. No one would question the generalization that herbivore biomass per unit area is lower in the north temperate forests than in a tropical rainforest, and that this difference is food based; but it is proposed that population abundance of a given *sexually* reproducing herbivore species in the less productive environment does not have to be food limited. Spatial constraints involving population persistence via sufficient encounter of individuals may ultimately be limiting to abundance. In this sense there may be, not an "enigma of balance," but faulty logic.

In other words, the manner in which living genomes ensure free-crossing, in relation to spatial constraints and behavior, may limit population abundance without the requirement for competition for food. Energy flowing through an ecosystem in this view may have little to do with regulation of population abundances of sexually reproducing species; and it is the regulation of abundance that is central to the mechanism of evolution.

···10···

ECOLOGICAL COMPONENTS OF EVOLUTIONARY THEORY

The member/vagrant hypothesis, and the evidence that supports it, infers that competition for resources is frequently not critical to population regulation in the oceans. If this is accurate, what are the implications for evolutionary theory? The remainder of the essay deals with this question.

To evaluate the potential significance to evolutionary theory of a change in perspective on population regulation in the oceans, it is necessary to clearly define the present role of population regulation in the evolutionary synthesis. The concepts are subtle, and thus may be more clearly presented within a historical sketch that traces the historical development of the natural history or ecological components of evolutionary theory from Darwin to the synthesis. For a detailed review of the development of evolutionary concepts within the synthesis itself, see Eldredge (1985).

Changing Role of Natural Selection

Because Darwin had to refute the theory of special creation of species, certain of the stronger arguments in support of creationism had to be given particular emphasis. The features of species that supported the creation theory were:

- the sterility of hybrids between species in contrast to the relative vigor of mongrels produced from the mating of varieties;

- the discontinuous geographic distributions of species and genera (which favored the multiple creation concept); and

- the lack of intermediate forms in the fossil record.

Darwin in his proof of evolution, *The Origin of Species* (1859), addressed these three points head on. He downplayed the reality of the species as a special level within systematics (including the importance of the sterility/fertility discontinuity between species and varieties) and upgraded the importance of varieties; he emphasized the powers of dispersal and migration of species (to counter the multiple creation concept); and he questioned the completeness of the fossil record. The strength of his rebuttal, although necessary in 1859 to refute the creationist theory, may have had a constraining influence on subsequent developments in systematics, biogeography, ecology, and paleontology.

Darwin took a two-pronged approach toward the first point. On the one hand, he emphasized the continuity between varieties and species:

> I look at the term species, as one arbitrarily given for the sake of convenience to a set of individuals closely resembling each other, and that it does not essentially differ from the term variety, which is given to less distinct and more fluctuating forms. (p. 108)

On the other, he de-emphasized the importance of hybrid sterility between species:

> In all other respects, excluding fertility, there is a close general resemblance between hybrids and mongrels. Finally, then, the facts briefly given in this chapter do not seem to me opposed to, but even rather to support, the view that there is no fundamental distinction between species and varieties. (p. 290)

Darwin's hypothesis on the mechanism of evolution, natural selection, does not deal with the origin of reproductive isolation (perhaps the key characteristic of speciation). Natural selection, through competition for limited resources, results in survival of the organisms that are better adapted to their environment, but as a concept it does not address the origin of reproductive isolation itself.

In the overall argument in *The Origin of Species* this weakness is not pointed out by Darwin, but it has been the focus of much of the subsequent debate in the evolutionary literature. For example, Fisher (1958b) discusses Darwin's analysis of the origin of diversity due to natural selection. He states that Darwin "is leading towards the problem of the diversification of species within a genus, which has more recently been called 'speciation,' and he does not attempt any detailed discussion of how the fission of a single interbreeding population into two can be brought about" (p. 291).

Mayr (1957b) draws a similar conclusion, and infers that the root of the problem was Darwin's concept of species themselves. Darwin, he says, "had fallen down in his attempt to explain the multiplication of species because he was not fully aware of the multiple aspects of species. Nor did he realize that multiplication of species, and origin of discontinuities, is not the same as simple evolutionary change" (p. 385). Mayr (1957a) goes further: "Having thus eliminated the species as a concrete unit of nature, Darwin had also neatly eliminated the problem of the multiplication of species. This explains why he made no effort in his classical work to solve the problem of speciation" (p. 4).

Two threads can subsequently be traced. First, systematists actively studied population patterns and richness in relation to the weakened species concept of Darwin. Second, the nature of reproductive isolation between species was described in considerable detail. The generalizations that emerged are paradoxical. In the search for the support of Darwinism, the naturalists reinforced the species concept. Their accumulated observations led to the *Rassenkreis* and the biological species concept of the new systematics, that species are groups of relatively discrete self-sustaining populations. Reproductive isolation rather than morphological discontinuities became the critical distinguishing feature of sexually reproducing species. The downplaying of the species category was reversed, and the importance of reproductive isolation (a broader concept than hybrid sterility) was emphasized. What was retained was the identification of population richness as evidence of speciation in action or micro-evolution.

This shift to the biological species concept itself (including the emphasis on the origin of reproductive isolation) disturbed the overall balance of the interactive arguments of *The Origin of Species*. Natural selection with a stripped-down species concept works. The emphasis is on morphological adaptation. However, with the enriched species concept defined within the new systematics in relation to reproductive isolation, the emphasis shifted to include speciation. Natural selection was elevated to do a job in the evolutionary synthesis that it did not have to do in either Darwinism or neo-Darwinism. From a careful reading of Dobzhansky (1937) and Mayr (1942) it is clear that the mechanism of the origin of reproductive isolation, and the putative role of natural selection in this aspect of speciation, was not satisfactorily resolved.

Dobzhansky (1937, chaps. 8 and 9) treats the issue of the origins of reproductive isolation and hybrid sterility. Most of his discussion deals with how reproductive isolation is maintained. The mechanism of the origin of isolation itself, the essence of speciation, is dealt with rather briefly and with some difficulty. Dobzhansky recognized that the role of natural selection in the origin of the isolation mechanisms *per se* was open to question:

> The spread within a population of genes that may eventually induce isolation between populations is probably due to their properties other than those concerned with isolation. What these properties are is a moot question, and here is the weakest point of the whole theory. (p. 258)

He states that genes which strengthen reproductive isolation "may be favored by natural selection" (p. 258).

Mayr (1942) also squarely faces the question of the mechanism of the origin of reproductive isolation in his chapter on the biology of speciation:

> A species is defined by us as a reproductively isolated group of populations, and it is obvious from this definition that species must possess isolating mechanisms which safeguard this reproductive isolation. . . . The question is: What are these biological barriers and how do they originate? (p. 247)

Mayr covers much of the same ground as Dobzhansky in describing what the barriers are but sidesteps how they originate. In a short section on selective factors and species formation (pp. 270–273) purportedly addressing this key question, he states:

> We are merely concerned with the question of how selective factors influence the establishment of discontinuities and what selective factors tend to enlarge the gaps between incipient species. Competition and predation are generally listed as the two most important factors to be considered in this connection. A survey of this field indicates, unfortunately, that our knowledge of the actual influence of these factors on the speciation process is still very slight. In fact it is surprising how badly ecologists have neglected these questions. (p. 270)

His subsequent review of the evidence does not clarify the role of natural selection in the origin of reproductive isolation (i.e., of speciation).

Stebbins (1950) was perhaps the most explicit of the principal contributors to the evolutionary synthesis in distinguishing between the different processes involved in adaptation and speciation: "The processes responsible for evolutionary divergence may be entirely different in character and genetically independent of those which produce isolation mechanisms and consequently distinct species" (p. 190). This conclusion appears later in the book in stronger terms: "Descent with modification and the origin of species are essentially different processes" (p. 249). The ecological aspects of the mechanism of the origin of reproductive isolation are not, however, well understood.

The recent literature indicates that this issue, i.e., the role of natural selection and adaptation in the origin of reproductive isolation, is still a point of contention. Ringo et al. (1985), for example, have taken an experimental approach using *Drosophila* sp. to evaluate the role of adaptation in speciation. The experiment is a test of Carson's "founder-flush-crash" hypothesis of speciation. The study is not conclusive, but they state, "The initial establishment of reproductive isolation occurs easily as a by-product of adaptive processes; we concur with Templeton that adaptation is still alive and well and playing a critical role in the origin of species" (p. 675). However, they also state that the process leads to a plateau and infer that reproductive isolation may not be completed by micro-evolution. The study, a useful introduction to the more recent literature on the mechanisms of the origin of reproductive isolation (i.e., speciation), focuses attention on perhaps the most critical unresolved problem in evolutionary theory.

In sum, Darwin's treatment of the species concept itself (including the sterility/fertility discontinuity) generated surprising developments. These include a strengthened species concept based on reproductive isolation, a larger role for natural selection (i.e., in the origin of reproductive isolation as well as adaptation), and an emphasis on geographic isolation.

Darwin's treatment of the other two areas of evidence in support of creation of species (discontinuous distributions and the lack of gradualism in the fossil record) has had equally strong influences on subsequent research. In ecology and biogeography there has been a marked emphasis on dispersal. The relative importance of dispersal in explaining biogeographical patterns is now at the heart of a major split within this discipline. We argue here that in ecology the emphasis on dispersal, particularly in relation to marine fish, has inhibited the understanding of population regulation. The present lively debate within paleobiology on the reality of stasis in the fossil record (a punctuated equilibrium versus gradualist interpretation of the fossil record) is a direct result of Darwin's treatment. Darwin's "proof" of evolution and his effective rebuttal of creationism have left a dramatic legacy. The weakened species concept was turned around quickly; the discontinuity in the fossil record is being resolved at present.

This essay addresses the ecological aspects of the generation of reproductive isolation and the importance of dispersal. They do not appear to have been satisfactorily resolved, and represent perhaps the final aspects of the legacy of what could be called Darwin's overkill. All of these aspects of his rebuttal of the evidence for special multiple creation (the species concept, generation of reproductive isolation, importance of dispersal, and continuity/discontinuity in the fossil record) should eventually receive a more balanced treatment than is given in the evolutionary synthesis. In its present conception, however, natural selection has been assigned a larger role in evolution than Darwin gave it—i.e., as the mechanism of speciation within a strengthened species concept.

Ecological Basis of Natural Selection

Natural selection, the mechanism of evolution, was deduced by Darwin and Wallace from two observations:

- relative temporal stability in the abundance of populations;

- high potential rate of population increase due to excess fecundity.

From these coupled observations they deduced that there must be intense intraspecific competition for resources (or in the avoidance of predators). Van Valen (1982) has structured Darwin's argument in the form of deductive logic. Because he clearly identifies the role of competition for limited resources in his analysis, the structure of the argument is quoted in full:

D1. L is the amount of resource R available.

D2. M is some minimum amount of resource R (see A1).

D3. N is the maximum stable number of population W, as determined by L.

D4. Struggle for life occurs in population W when even at least one member of W dies because of having an insufficient amount of a resource and this lack is due to the size of W being greater than N.

A1. If any individual in population W does not receive at least some minimum amount M of some resource R, this individual will die because of having an insufficient amount of a resource.

A2. At any finite size of a population W, the number of next-

generation offspring produced is greater than the size of
the parental population.

A3. The size of W will increase until it exceeds N.

A4. Whenever the size of W is greater than N, one or more
individuals in W do not receive at least M.

A5. N is finite or 0.

C1. The number of offspring produced in W will sometimes
be greater than N (A2, A3, A5).

C2. One or more individuals in W will sometimes not receive
at least M (C1, A4).

C3. Struggle for life occurs in W (C1, C2, A1, A4, D4).

Extension of the argument to natural selection involves assumptions on variation and fitness; extension to genotypic
change involves heritability. (pp. 328–329)

The above analysis of the deductive structure of natural selection as developed by Darwin (and Wallace) allows one to precisely
identify the ecological component of natural selection. About the
controversial role of competition in population regulation, Van Valen
states, "Probably most ecologists on both sides of the perennial controversy on the frequency and importance of competition will accept A3 as
a touchstone" (p. 329). But A3, which says that the size of a population
will increase until it exceeds its maximum stable abundance (which is
somewhat tautological), is dependent on D3 and D1, which say that the
maximum stable abundance is determined by available resources. It is
precisely this issue that we are disagreeing with in the member/vagrant
hypothesis: contrary to D3, we do not think that N is necessarily
determined by L.

The critical assumption, then, is that population abundance is
defined ultimately as a function of limited resources. For geographically constrained sexually reproducing populations in the oceans the
logic is flawed: the maximum stable abundance of a population is not
necessarily determined by available resources. In this sense intense
intraspecific competition is not the only inference that can be drawn
from the coupled empirical observations of temporal stability in numbers and high reproductive output. As developed in the first half of this
essay, spatial processes in relation to animal behavior may limit the
abundance of animals in the appropriate geographic area of population
distribution.

A second ecological aspect of natural selection (i.e., one in
addition to the assumed critical importance of competition in popula-

tion regulation) is Darwin's associated treatment of absolute abundance and the amount of individual variability: "It is the common, the widely diffused, and widely-ranging species belonging to the larger genera, which vary most, and these will tend to transmit to their modified offspring that superiority which now makes them dominant in their own countries" (p. 170). In essence Darwin argues that natural selection results in increased abundance, giving the more abundant species the greater range of variability. Fitness as manifested at the population level is increased abundance. Because Darwin believed that abundance is a reflection of natural selection, he was not sympathetic to the argument that the more important events in evolution occur in small, peripherally isolated populations. This second ecological aspect of natural selection thus has three components: population richness is a function of abundance; innovation in evolution predominantly occurs within the central core of the distribution of the species; and absolute abundance is generated by natural selection.

These three components have been substantially modified in the evolutionary synthesis. First, the accumulated studies of the population systematists do not demonstrate that population richness is a function of the overall abundance of individuals in a species. Mayr (1942, p. 236) makes this point: "The often-made statement that 'abundant species are more variable than rare ones' is not necessarily correct. Some very rare polytypic species are much more variable than some common monotypic ones." Second, the synthesis gives peripherally isolated populations a major rather than a minor role in the evolution of novelties.

The third point, absolute abundance, is treated inconsistently in the synthesis. The Fisher school of mathematical population genetics, which emphasizes the function of adaptation by natural selection, is strictly Darwinian; it believes that the result of natural selection is increased abundance of population and of the species. In Fisher's fundamental theorem of natural selection, which forms the basis of much of population genetics, absolute abundance is treated very much in a Darwinian manner. For example, in the second edition of his 1930 contribution to the synthesis, Fisher analyzes the results of natural selection on the Malthusian parameter (1958a, p. 45):

> The balance left over when from the rate of increase in the mean value of M [the Malthusian parameter] produced by Natural Selection is deducted the rate of decrease due to deterioration

in environment, results not in an increase in the average value of M, for this average value cannot greatly exceed zero, but principally in a steady increase in population.

In other words,

> Any net advantage gained by an organism will be conserved in the form of an increase in population, rather than in an increase in the average Malthusian parameter, which is kept by this adjustment always near to zero.

It is important to note that "population" in his definition is the total number of individuals of the species. Thus, natural selection acts by increasing the fitness of individuals, and this leads to increased abundance, which is essentially identical to Darwin's concept of natural selection. Other contributors to the synthesis (Dobzhansky 1937, Mayr 1942, Wright 1931) do not act as if this is really what is happening as a result of natural selection.

In sum, natural selection plays a different role in the evolutionary synthesis from what Darwin gave it because of the changing species concept, which has expanded to include the generation of reproductive isolation during speciation. The associated ecological aspects of natural selection (absolute abundance, population richness, and the relative importance of central versus peripheral parts of the distribution) have also been significantly modified. Perhaps the only constant has been the putative role of intraspecific competition as the backbone of natural selection. The role of natural selection in evolution has changed substantially since 1859, but its underlying mechanism (survival of the fitter animals through competition) has remained constant. It is the evidence for the importance of this mechanism in the oceans that we are questioning in this essay.

Problems with the Evolutionary Synthesis

Before considering life-cycle selection and its putative role in speciation in the next chapter, it is useful to summarize some problems with the evolutionary synthesis. It is noteworthy how rapidly points of contention are currently being identified, given the somewhat deceptive finality of the post-synthesis discussions. Mayr (1980), speaking of paradigm shifts, states that the evolutionary synthesis was not "still another revolution but simply the *final* implementation of the Darwinian revolution" (p. 43; emphasis added). The ink was hardly dry on that article (which was probably written in 1974 although not published until 1980) before doubts surfaced about the sufficiency of the synthesis to account for all evolutionary phenomena.

One sees a parallel here. Vigorous research by naturalists in support of Darwin's inference that populations are incipient species led first to a strengthened biological species concept defined in relation to reproductive isolation and then to dissatisfaction with natural selection (Gulick 1890; Heincke 1898; Kyle 1900). Today, vigorous research on a broad front in support of the evolutionary synthesis is generating anomalies that require considerable footwork by the synthesists for the anomalies to be accommodated under their umbrella (see Stebbins and Ayala 1985 for a recent review). A listing of the characteristics of species provides a framework for highlighting some of these problem areas within the synthesis.

To begin with the obvious, it is species, and thus their defining features, that are the product of speciation. The key features of the biological species concept are:

1. reproductive isolation
2. frequent sterility of hybrids
3. component population richness (from one to many)
4. mean number of individuals
5. morphology
6. duration

The first four are characteristics considered on *ecological* time scales. Adaptive components of morphology itself, with the exception of morphological features that ensure reproductive isolation, are not a necessary result of speciation events. For paleobiologists, however, speciation events can be traced only to the degree that there are morphological changes. Morphology and duration, therefore, are characteristics of species as studied on *geological* time scales. On both time scales there is a geographical frame of reference for the distribution of the species; but in general, with a few notable exceptions (for example, Coope 1979 for Coleoptera), geographical changes in component *population* structure cannot be described on geological time scales.

These, then, are the six features that a theory of speciation should account for within a geographical framework on the two time scales of study. As already indicated, the evolutionary synthesis, like Darwinism, still has difficulty with reproductive isolation and sterility of hybrids. They are not directly selected for or against, according to the synthesis; that is, they are not a result of adaptation. The generation of reproductive isolation and hybrid sterility by natural selection is not adequately accounted for by, for example, Dobzhansky (1937) or Mayr (1942) (see the first section of this chapter). These aspects of species, and thus speciation itself, are considered in the synthesis as by-products or incidental results of natural selection of the fitter animals in relation to particular environmental requirements.

It is perhaps not surprising that the sexual aspects of speciation are fuzzy in the evolutionary synthesis, since the constraint of sexual reproduction itself is not satisfactorily understood in spite of a rich literature. Darwin (1859), in a discussion of hermaphrodites, recognized that the essential need for sexual reproduction between individuals "is a general law of nature (utterly ignorant though we be of the meaning of the law)" (p. 143). Mayr (1982a) has identified the interpretation of the near ubiquity of the sexual mode of reproduction as one of the important unresolved questions. This in itself, in view of the fact that the key feature of speciation is reproductive isolation, has to be considered a problem with the synthesis.

The third characteristic, population richness, even though a key component of the synthesis, is also not handled adequately. There has been limited explanatory power relative to species-specific differences in population richness. It bothered both Mayr and Rensch that some species are not polytypic, yet the interspecific differences in population richness have not been explained.

Descriptions of the very existence of population richness for most species were the substantive contribution to the synthesis by the systematists and ecologists (natural historians at the time), yet the overall theory (and subsequent developments) has not persuasively accounted for the between-species differences. Mayr (1942), in the section on population richness, states, "The conclusion from this evidence is that a high proportion of the species in the well-worked animal groups have been found to be polytypic but that a certain percentage of species does not break up into distinct geographic races." In other words, many species such as checkerspot butterfly and Atlantic herring are composed of many populations, yet some species comprise a single or only a few populations, such as the satyrnine butterfly and the European eel. Since population structure was considered evidence of speciation in action, species without such patterns were thought to be somewhat anomalous. Rensch (1959) asks the "why" question: "Considering the geographic race formation as the most frequent and, in many groups, quite usual precursor of species, we still have to ask whether recent animal forms without geographic variation... have passed stages of a *different* kind. A sure answer can rarely be given" (Rensch's italics). Rensch, by referring to "recent animal forms without geographic variation," is effectively inferring that such species as satyrnine butterfly and European eel are more recently evolved than checkerspot butterfly and Atlantic herring.

More recently, Carson, in his paper on the genetics of speciation at the diploid level (1975), addresses the same question, i.e., the reason for the difference between panmictic and polytypic species that Mayr did not address and Rensch could not explain. He argues that this difference in the fundamental characteristic of diverse species is due to the inferred existence of two systems of genetic variability (open and closed systems): "The difference between polytypic and monotypic species may in a large measure be because the monotypic species lacks an open-variability system or has an insignificant one." Carson in this way argues that geographic population pattern and richness structure are a function of between-species differences in genetic constraints. It is clear that the evolutionary synthesis has not accounted for the empirical observations on species-specific differences in population richness. As indicated in Chapter 5, the member/vagrant hypothesis accounts for this third characteristic of species in strictly ecological terms.

The fourth characteristic identified, the mean number of individuals in a species, is addressed next. The role of absolute abundance of animals in a population is difficult to grasp in the evolutionary synthesis. As summarized in the previous section, Darwin (1859) believed that absolute abundance or commonness was of key importance in the evolution of species by natural selection, even though it was unrelated to fecundity itself. He considered abundance to be an indicator of competitive success.

Subsequent developments have treated abundance in radically different ways. Its treatment falls into two categories depending on the relative emphasis the scientist places on adaptation and speciation. Fisher, in the development of his fundamental theorum of natural selection, was interested in adaptation, not speciation. In his mathematical treatment, absolute abundance of populations reflects the result of increases in fitness due to natural selection. He stresses the importance of large populations and defines N in his mathematical development of the genetical theory of natural selection as the total number of individuals of the species. Only large numbers can incorporate into the gene pool the mutations of extremely low mutation rate that he considers essential for phyletic evolution (or adaptation).

Wright (1931), in contrast, is particularly interested in speciation. He deals with small populations, and N in his mathematical development is the number of individuals in the population that can effectively, rather than potentially, interbreed, clearly a much smaller number than Fisher's. Wright, in his "three-phase shifting balance" theory, proposes that populations are sufficiently small that random drift is important in determining gene frequencies. It was this disagreement on the definition of N that, as much as anything, led to the falling out of these two founders of theoretical population genetics (see Bennett 1983, pp. 40–48).

It is not differences in mathematics that separate the two schools of theoretical population genetics, but rather emphasis on two different aspects of evolution (adaptation and speciation), and thus differences in their underlying assumptions. However, it should not be overlooked that even in the subject matter of common interest very different conclusions were reached. Wright (1934), for example, concluded that "adaptative advance [as well as speciation] was attributed more to intergroup than intragroup selection." Fisher found Wright's arguments fallacious. The theories developed could hardly be more different. Fisher relies solely on individual selection, Wright on a form of group selection. Fisher emphasizes the importance of large populations for adaptive modification; Wright believes that small populations are more important for this process. Fisher stresses the importance of natural selection; Wright stresses the importance of random genetic drift. Fisher emphasizes outbreeding; Wright emphasizes inbreeding.

These differences, which focus on the relative importance of population abundance in evolutionary processes, are sustained within the evolutionary synthesis. Simpson stresses the importance of phyletic evolution (see Gould 1980 for an analysis) and thus implicitly follows Fisher's school of population genetics. Mayr (1954) stresses the importance of small, geographically isolated populations and thus fits well into the Wright school. Lewontin (1980) analyzes the differences between the two approaches and concludes that evaluating their relative value is very difficult: "Competing hypotheses must be judged on the basis of a guess about the product of a relatively large but unknown number, the effective size of a population, and a relatively small but unknown number, the migration rate, or the selection coefficient, or the mutation rate for genes, which makes it extremely difficult to distinguish between various hypotheses" (p. 62).

Given the radical differences between the two approaches, and the key role of the mathematical models of the theoretical population geneticists in the development of the evolutionary synthesis, it is pertinent to question how satisfactory the synthesis really is, particularly as it relates to the central issue of the role of absolute abundance in adaptation and speciation. Part of the reason for the differences in treatment of the role of absolute abundance seems to have been the necessary coupling of adaptation and speciation, which is required if intraspecific competition for resources is the mechanism for the origin of species (i.e., the origin of reproductive isolation). However, it is also to be noted that neither approach within the synthesis (the big N versus small N schools) generates a hypothesis to account for one of the key characteristics of species, the number of individuals (N itself). This is a generally unstated limitation of the synthesis. Each of the two schools of population genetics within the evolutionary synthesis treats commonness in a different manner, yet neither approach accounts for empirically observed species-specific absolute abundances.

The fifth characteristic of species, morphology, has been the focus of much of the attention in evolutionary theory. Perhaps a major problem in the treatment of morphology in the evolutionary synthesis is its failure to explain why speciation does not necessarily involve substantive differences of form. Why does speciation sometimes involve such minimal morphological adaptation (other than features associated with reproductive isolation)? The very observation that speciation has occurred without such adaptation is a strain on the gradualist theme of the synthesis. The copepod species of the *Labidocera jollae* group (Fleminger 1967) and the European/American eel species are good examples of limited morphological adaptation during speciation. In addition, substantive morphological differences can occur between populations of the same species (see Hedgecock 1986 and Palmer 1985 for a review of this literature). Morphological adaptation and speciation can be decoupled.

The difficulty emphasized here is the explanation of speciation with limited morphological adaptation, rather than the accounting for the adaptive value of particular morphological features (which has been criticized by Bateson [1909] and Gould and Lewontin [1979], both papers using the metaphor of Pangloss). In addition, saltations in form seem required from the accumulated evidence (as, for example, during paedomorphic events). Yet, within the synthesis the inferred existence of such saltations sits uncomfortably. Neither the relative lack of morphological change during some speciation events nor the inferred requirement for substantial change in form in the historical record is readily accommodated.

The sixth characteristic listed, duration of species, has been the subject of intense discussion during the past decade. Under gradualism one would not expect to observe what Simpson (1944) has defined as the bradytelic mode of speciation, which includes the existence of "immortal" forms. Stasis itself may be considered an unexpected observation within the synthesis. Dover (1985), in response to a book review by Dawkins, summarizes this issue succinctly:

> If we accept, for the moment, that stasis is a widespread observation in the geological record, and if too we accept that the forces behind selection are the only means for promoting biologically useful novelties throughout a population . . . , then it is surprising that little significant evolution occurs over periods of time that may span from hundreds of thousands to millions of generations, in a diverse range of species. The point is not that evolution is gradual "*during the times that it is actually happening*" (Dawkins' emphasis) but that it seemingly is not happening for most of the time. This was the message of the infamous Chicago meeting, which many of us took home, and which demands an explanation. (p. 19)

In the remainder of his letter Dover briefly reviews how stasis has been incorporated into the evolutionary synthesis. The arguments, as he points out, are not convincing. Stasis in the face of natural selection as the mechanism of evolution demands an explanation of an *ecological* nature. In addition, the temporal pattern of extinction in the oceans has been difficult to reconcile with natural selection as the principal generating force.

In sum, the evolutionary synthesis has to strain to account for the fundamental characteristics of species, as well as speciation and extinction itself. This perceived strain may be a product of the very nature of the synthesis. It was an attempt to patch up Darwinism, which itself did not treat species and speciation clearly, the emphasis having been on the proof of evolution itself and on the generation of morphological adaptations. This is not to say that the synthesis does not treat the six features listed above, but rather that the explanatory power in relation to these species' characteristics is limited in some cases and strained in others.

Four additional points are worth noting, the last two taken from Stebbins and Ayala (1985). First, the present heavy emphasis on the importance of small, peripherally isolated populations in evolution is restrictive. Second, the key role of intraspecific competition for limited resources (relative to population regulation) in natural selection, as clearly analyzed by Van Valen (1982), lacks empirical support in the ecological literature. Third, the degree of genetic heterozygosity is much larger than expected. Finally, molecular evolution is observed to be relatively constant (the molecular clock concept).

Concluding Remarks

In the first section of this chapter it is pointed out that Darwin did not adequately address the concept of speciation in the *Origin*. He avoided the issue by defining species themselves in a particular manner. Thus, natural selection involved only adaptation; and evolution in essence was considered to be adaptation. Subsequent field studies by natural historians and systematists resulted in a strengthened species concept involving reproductive isolation between groups of populations (i.e., the biological species concept of the new systematics). Part of the synthesis involved the incorporation of speciation (the origin of reproductive isolation) into evolutionary theory. Evolution is considered in the synthesis to comprise the processes underlying both adaptation and speciation. Natural selection, as a result, plays a larger role in the evolutionary synthesis than in Darwinism. Reproductive isolation, as well as adaptation, is selected for. Also, geographic isolation barriers are considered to play a major role in the process of speciation.

In the second section of the chapter the *ecological* components of natural selection are briefly summarized. Natural selection in Darwinism involves population regulation. Intraspecific competition for limited resources, and avoidance of predation, generates natural selection (given individual variability within populations and a heritable component for the variability in traits). Natural selection is a deduction based on the empirical observations of stable abundance and high fecundity. The ecological conceptualization of natural selection as defined by Darwin has not substantially changed within the evolutionary synthesis. The role of absolute abundance and small populations does, however, receive a different emphasis in the synthesis from what Darwinism gives it.

In the third section of the chapter, it is argued that the characteristics of species, or the results of speciation, are not well accounted for within the synthesis. The argument, in essence, infers that the theory dealing with the process of speciation should provide explanatory power for the fundamental characteristics of species. The synthesis is considered to be strained in this respect.

The chapter, if only because of its length, is undoubtedly a simplistic summary of a complex literature. It is hoped, however, that the overview is a balanced one, and that it provides an appropriate perspective for reconsidering the role of population regulation in speciation.

···11···

LIFE-CYCLE SELECTION AND SPECIATION

In this chapter we explore an alternate or enlarged view of population regulation, the member/vagrant hypothesis, which de-emphasizes intraspecific competition for resources, and its significance for the origin of reproductive isolation (i.e., for speciation). Subsequently, the six characteristics of species defined in Chapter 10 are reconsidered within this alternate view. A modest reconceptualization of the selective process underlying speciation, a process involving life-cycle as well as food-chain selection, is argued to provide increased explanatory power relative to these fundamental characteristics of species. Further, contemporaneous components of population regulation are argued to be associated with speciation and energetics adaptation.

Life-Cycle Selection Concept

Four constraints are involved in speciation of sexually reproducing animals:

1. Information for metabolism and development is chemically mediated and is in a discrete rather than a continuous form; thus, there is particulate inheritance of information across generations.

2. Biochemical information degrades with time. The replication of information is not perfect.

3. Sexual reproduction conserves information content and may retard or mask the errors in its replication (Bernstein et al. 1985).

4. Nonequilibrium conditions are generated by changes in the geographical or spatial framework underlying the relational phenomena between similar individuals (which is demanded by the constraint of sexual reproduction).

From the starting point of these constraints, speciation is perceived to be the result of trying not to change in the face of dissipation of information in the biochemical hereditary material and a shifting spatial matrix within which populations are striving to persist. The spatial matrix includes the physical geography as well as the structure provided by other species of plants and animals. In the member/vagrant hypothesis, life cycles and population patterns of sexually reproducing animals are defined in relation to particular geographical or spatial constraints that ensure persistence.

Life-cycle selection is a logical extension of the member/vagrant hypothesis on population regulation, just as the originally defined concept of natural selection is an extension of the Malthusian doctrine and the hypothesis on population regulation through intraspecific competition for limited resources (see Chapter 10). The member/vagrant hypothesis focuses on the losses from populations during cyclical sexual reproduction in a fixed spatial context; life-cycle selection focuses on the changes in frequency distribution of traits in populations generated by membership versus vagrancy during life-cycle closure. In essence, what is selected for is the ability to continue to be a member of a spatially defined population. The emphasis is on those aspects of the life cycle as a whole that ensure population persistence in relation to the spatial dimension. Life-cycle selection is considered to encompass sexual selection and mate selection, which deal with reproductive processes at the adult stage of the life cycle (ensuring coupling of mates in particular geographical space). It also encompasses distributional and behavioral phenomena at other stages of the life history. It acts at the individual level. Life-cycle selection, we argue, contributes to speciation in the oceans with or without geographic barriers.

The coupled observations of relative stability of numbers of individuals of a particular species and high reproductive output led Darwin to infer or deduce the existence of intense competition for limited resources between individuals of the same species. This inference is the basis of natural selection. It has been argued here that the inference or deduction was partially flawed in the sense that in the oceans there is considerable evidence that food chain interactions are not necessarily critical to population regulation. A portion of the accumulated life-cycle losses may be due to intraspecific competition for resources and predation, but it is argued to be a small portion for many species with complex life histories.

This very minor shift in emphasis from food-chain constraints to space may have subtle but important implications. Reproductive continuity and its opposite, reproductive isolation, are the very focus of both the member/vagrant hypothesis of population regulation and the life-cycle selection concept of speciation. Reproductive isolation with life-cycle selection is not an incidental by-product of energetics adaptations of other aspects of the life history; it is the very heart of the mechanism of population regulation and life-cycle selection. The reproductive isolation between populations that is associated with philopatry may be generated without the necessity of geographic barriers. Mobile animals of different populations can share a common distributional area for large parts of the life cycle, yet they manage to home to their natal spawning areas for sexual reproduction. On ecological time scales, the empirical evidence is overwhelming that geographic barriers are not required for reproductive isolation. The animals must know which populations they belong to when they sort themselves into spawning groups. The empirical observations on relative reproductive isolation between populations without geographic barriers suggest that speciation may not require geographic isolation. Successful coupling of members with vagrants, however, needs to be selected *against* for this to be the case.

Life-cycle selection is considered a conservative process generating persistence of populations of sexually reproducing animals. Two creative processes are defined: the internal dissipation of the hereditary material at the molecular level by mutation and copying errors, and the external changes in the physical geography or spatial fabric of the environment (including the structures provided by other species of plants and animals). The first process at the molecular level may generate reduced viability of the offspring of crossings between individuals of different populations. In such cases reproduction between individuals of separate populations is highly selected against. This is conventional wisdom for several modes of speciation. What is added here is the member/vagrant mechanism of population regulation. Life-cycle selection, by continuing to select for membership in specific populations, acts without geographic barriers to reinforce any biochemical incompatibility that arises between individuals of separate populations.

Speciation events in this internally generated mode are thus ultimately a function of "neutral" genetic or molecular divergence. Speciation could occur without adaptation, other than changes associated with reproductive isolation. Life-cycle selection only reinforces a biochemical internal forcing function. The European eel and the

American eel, for example, have speciated with very little morphological divergence. According to Schmidt (1922, pp. 203–205), the differences between the two species involve life-cycle timing and physical geography: "Despite the fact that the two species are outwardly so alike as to be hardly distinguishable, they differ to such an extent that the one takes about three times as long as the other to pass through the same cycle of development."

Frost and Fleminger (1968) note the relative lack of difference between sympatric species of *Clausocalanus:*

> Our experience with *Clausocalanus* and the findings of others as discussed above can be summarized as follows:
>
> 1. Most of the differences between species of *Clausocalanus* are found in primary and secondary sexual structures.
> 2. Many species of *Clausocalanus*, both within and between groups, co-occur extensively.
> 3. No appreciable differences are found in the oral appendages of *Clausocalanus*.
> 4. Only minor differences in selective feeding have been demonstrated among sympatric species of other copepod genera.
>
> Accordingly, while there is ample evidence of selection operating on primary and secondary sexual characters among sympatric species, there is no obvious indication of selection acting to separate these species with respect to the food niche. (p. 89)

Frost and Fleminger do, however, note that the morphological characters involved in speciation within this genus involve relational phenomena between individuals that are consistent with the member/vagrant hypothesis of population regulation: "Differences between inter- and intragroup characters are heavily concentrated in structures strongly influenced by sex, but suggest two independent sequences of evolution, both presumably in relation to *wastage of gametes*" (p. 88; emphasis added).

The second creative process is externally forced and as a result may have a more important ecological component. Under the member/vagrant hypothesis, life cycles (both behavioral and morphological aspects) permit population persistence within particular spatial constraints. Differential changes in the spatial fabric to which life cycles are associated in part of the distributional range of the species may have the potential to break the cycle of populations. It is these putative life-cycle-breaking events that may generate the more saltational speciation events through paedomorphosis.

In the summary of evidence in support of the member/vagrant hypothesis (Chapter 8), we identified an overall secular trend within the evolution of sexually reproducing animal species toward the collapse of complex life cycles. This trend has corresponded to an internalization of the early life-history stages and a concomitant independence of the life cycle from geographical constraints; that is, the dispersal of early life-history stages has been minimized by their gradual inclusion within the embryo.

Superimposed on the general trend toward a simplification or internalization of life cycles (see Strathmann 1985 for a recent review) is evidence, particularly in the marine environment, of considerable shifting back and forth in the process, principally through paedomorphosis, i.e., the retention of ancestral juvenile characters by later ontogenetic stages of descendants. There are two forms of paedomorphosis, neotony and progenesis. Neotony is paedomorphosis produced by retardation of somatic development, while progenesis is paedomorphosis produced by precocious sexual maturation of an organism still in a morphologically juvenile stage.

Hardy (1953) has called this repetitive process a "retreat from specialization." This process, which is often a change in the life cycle by modifications in animal development, can result in the saltations in form that are associated with mega-evolution. Again in Hardy's words, paedomorphosis is an element "of cardinal importance in the general process of evolution as especially in the field which G. G. Simpson would call mega-evolution" (p. 123). The ecological factors associated with these life-cycle changes had not at that time been elaborated on, although de Beer (1940, p. 121) recognized their possibility:

> It is probable that ecological factors were not inactive in favoring the evolution of groups by paedomorphosis. The correlation between paedomorphosis and increased power of evolution may be due to an effect to which attention was first called by Young. He showed that some environments in which organisms live may be regarded as "difficult" and others as "easy."

Difficult environments were thought to generate paedomorphosis, but the sense in which *difficult* and *easy* were used was not defined. It is suggested here that changes in the physical geography that influence or disrupt life-cycle closure may be the source of "difficult" environmental challenges underlying paedomorphosis.

A hindrance to understanding the ecological aspects of paedomorphosis in the oceans may have been the generalization that the early life-history stages for marine animals are for dispersal. In the perspective developed in this essay, we hypothesize to the contrary, that these stages are associated with population persistence at particular spatial scales in relation to the constraints of the physical geography. From this perspective, paedomorphosis can be viewed as the product of a life-cycle-breaking event. It is suggested that such life-cycle changes are externally forced in the oceans, by changes in the physical geography that disrupt the existent tight linkage between the morphological and behavioral cycles of the animal and the associated particular geography. Changes in the physical geography, in this view, have the potential to break life cycles. The selection process, once triggered by the changing physical geography, may involve alterations in the animals' development through paedomorphosis to ensure population persistence under the evolving physical regime.

Many of the examples of paedomorphosis involve a move from a coupled benthic/pelagic life cycle to a solely pelagic life cycle. For example, Robert Gurney (quoted in Hardy 1953) concluded that pelagic copepoda "have arisen by arrested development from a larval form having the general characters of the Decapod Protozoea" (i.e., pelagic copepods have arisen from benthic/pelagic decapods). Such dramatic life-cycle changes may be induced by changing physical geography (i.e., life-cycle-breaking events) under spatial selection. The missing information is the internal mechanism. What happens developmentally when the life cycle cannot be completed because of changing physical geography? The ecological component involving population persistence mechanisms, and the forcing function of changing geography, can be readily envisioned; but comprehending the mechanisms by which larval forms attain sexual maturity through modification of the developmental program perhaps demands a leap into the dark.

The image of speciation in the oceans is a repetitive contracting and expanding of complex life cycles in response to a nonequilibrium physical environment and an imperfect biochemical information storing system. The conservative functions of sexual reproduction (the free-crossing of Chetverikov) and of life-cycle selection, in accommodating or responding to such nonequilibrium conditions, result in speciation.

With the constraint of sexual reproduction, and the associated requirement for life-cycle closure by populations, geographic barriers are not
necessary for the process of speciation. The process of regulating
abundance itself, as formulated in the member/vagrant hypothesis,
involves the maintenance of reproductive continuity within populations as well as relative reproductive isolation between populations.

Spatial Changes in the Oceans

There is considerable evidence that ocean circulation and
coastal zone configuration have changed substantially over a range of
time scales. During the Mesozoic the changes were fairly tranquil, but
they became less so during the Cenozoic (i.e., within the last 65 million
years). Berggren and Hollister (1977) describe this period of unrest:
"Continued continental dispersal and climatic deterioration (and lati-
tudinal thermal heterogeneity) due to high latitude cooling (beginning
40 m.y. ago but particularly the last 10 m.y.), led to accelerated surface
and bottom water circulation, especially along western margins of
ocean basins and the development of erosion and redeposition as a
major sedimentary process." They summarize it as "commotion in the
ocean."

Even on the much shorter time scale, in evolutionary and
geological terms (of thousands of years), the large-scale thermohaline
ocean circulation, as well as the coastal zone and continental shelf
circulation and mixing features, has varied dramatically. During the
last 700,000 years (the Pleistocene epoch), climatic fluctuations and
their attendant glaciation cycles have caused the extent of the continental shelves and their associated circulation to oscillate on an approximately 100,000-year cycle. The fluctuations in meltwater input into
the world's oceans associated with the glaciation cycle have influenced
the vertical circulation in a nonequilibrium manner. The intermittent
slowing in vertical mixing in the oceans caused by the pulsed addition
of meltwater during deglaciation periods has been called the "Worthington effect," and support for its existence is provided by Berger and
Killingley (1982) and Berger et al. (1985).

The studies cited above illustrate, albeit somewhat arbitrarily,
the dynamic nature of the changes in the physical geography of the
oceans even over time periods that are short relative to evolutionary
processes. These changes appear to have had biological effects. Berger
et al. (1983) provide evidence for a drop in fertility in the oceans,
inferred to be a result of a weakening of the east-equatorial upwelling

systems, during the glacial-Holocene boundary (about 10,000 years ago). Cronin (1985) links enhanced speciation of marine ostracods to the closing of the Isthmus of Panama (3 to 4 million years ago), which restricted circulation. Cronin notes that speciation "seems to have occurred within the zoogeographic range of the ancestral species" rather than in peripherally isolated populations.

Jablonski et al. (1983) show that major evolutionary novelties or innovations in marine benthic environments have occurred in the nonequilibrium coastal zone environment rather than in the more stable deeper water environments. They conclude, however, that speciation has occurred more rapidly in the latter areas. Buzas and Culver (1984), evaluating speciation rate of benthic foraminifera on the Atlantic continental margin of North America, reach the opposite conclusion: "The highest rate of speciation and extinction occurs in these shallower waters. The data indicate that the rate of evolution of species capable of living in the more variable, shallower environments is higher than that of the species living in the deeper, supposedly more stable environments" (p. 329).

The contradiction may be resolved by the analysis of extinction patterns on continental shelves by Sepkoski (1987). He notes that even though whole communities exhibit increasing extinction in the offshore more stable benthic environment (which is consistent with Jablonski's observations of higher speciation rates in this environment), individual taxonomic classes have their highest extinction onshore (which is consistent with Buzas and Culver). Sepkoski demonstrates that the "offshore trend at the community level results from a concentration of genera in classes with low chracteristic extinction rates in nearshore environments." He concludes that the observations are consistent with the generalization that extinction rates are higher in more physically variable environments.

Bretsky and Klofak (1985), in an analysis of expansion of range of benthic marine invertebrates within epicontinental sea regimes (i.e., marine environments that covered the continents during the Paleozoic), conclude that "throughout the Early Phanerozoic benthic marine speciations occurred preferentially in marginal marine environments" (p. 1469). The inference again is that speciation and extinction rates are higher in the physically more variable environments. Zinmeister and Feldmann (1984), in an analysis of Eocene rocks from Antarctica, similarly conclude that several genera and classes of marine invertebrates (including molluscs, echinoderms, and arthropods) evolved in "high-stress, shallow-water environments" before they moved to deeper water at lower latitudes during the Cenozoic. In sum, even though the causes of variable speciation rate and the evolution of novelties are not well understood, it seems clear that speciation in the oceans occurs during periods of marked changes in the physical geography.

Valentine and Jablonski (1983), in a review of speciation in shallow seas, in particular on the continental shelves, also emphasize the key role in speciation of changing circulation and the provincialization of the continental shelves during the Cenozoic. They note, however, that the repeated changes in sea level during the Pleistocene did not necessarily enhance speciation or extinction rates: "The Pleistocene invertebrates along the linear eastern Pacific shelves have suffered relatively little extinction, and so far as can be told have undergone little speciation despite repeated sea-level changes" (p. 216). Cronin (1982) draws the same conclusion from analyzing the ostracod fossil record. In his study, circulation changes, not rising or falling sea level, are considered critical. The ability of species to adjust their geographical distribution during the Pleistocene by tracking the appropriate environment has been remarkably illustrated for terrestrial Coleoptera by Coope (1979), who states, "An extraordinary degree of specific stability coincided with a period in the earth's history characterized by numerous large-scale fluctuations in climate, a situation that leads one to expect rapid speciation and numerous extinctions" (p. 262). Perhaps if there is a displacement rather than an elimination of the geographical infrastructure for life-cycle continuity, changes in sea level alone do not enhance speciation/extinction processes for all marine groups.

Finally, Lipps (1970) provides the evidence that the extinction/ species-radiation events in marine plankton paralleled fundamental changes in ocean climate. During geologically warm periods the oceans lacked vertical stratification. Species diversity of microplankton groups was very low and was characterized by morphologically simple forms which (in the case of foraminifera, at least) do not use the vertical structure of the oceans for their reproductive cycle. During cooler periods there was considerably more vertical structure in the oceans, which, Lipps hypothesizes, enhances maintenance of species in particular geographic space. Lipps infers that increased niche space and decreased competition are important in the radiation of species during these cooler periods.

Following the member/vagrant hypothesis and its role in selection as discussed above, we would argue that it is the changing geographical structure itself which permits greater or lesser opportunities for life-cycle continuity for sexually reproducing species. A stratified ocean with associated vertical differences in the horizontal circulation provides opportunities for life-cycle continuity and thus population persistence that are not available in a vertically homogeneous pelagic environment. If it is accepted that competition for limited resources in the oceans frequently plays a minor role in population regulation, we would infer that it is less critical to the extinction/radiation process than Lipps hypothesizes.

In sum, there is considerable evidence, only a taste of which is given here, that the physical geography of the oceans has varied significantly through geological time, including alteration in the degree of thermal stratification of the world's oceans. From the present-day zooplankton observations discussed in Chapter 7 it has been argued that the differential circulation features associated with vertical stratification are "used" by diverse zooplankton forms (including the larvae of fish and benthic invertebrates) to ensure population persistence. It is inferred here, following Lipps (1970), that the changing circulation features, including the occasional dramatic changes in thermal stratification and associated current structure, can be life-cycle-breaking events that force what has been termed "nonequilibrium speciation events." In this conceptual model, marginal populations are considered no more important than central ones.

Concluding Remarks

It was inferred at the beginning of this chapter that the subtle shift in emphasis on the nature of population regulation in the oceans, involving life-cycle continuity in relation to physical geography, and the related concept of life-cycle selection, helps account for the six characteristics of species (and thus speciation/extinction processes). First, life-cycle selection, and the associated component of population regulation, acts *directly* on the reproductive process. Reproductive isolation and sterility of hybrids are the central foci of life-cycle selection and thus speciation. As such, the mechanism of the origin of reproductive isolation and associated hybrid sterility is a logical extension of the member/vagrant hypothesis of population regulation. If the sexual mode of reproduction does indeed function to mask imperfections in the replication of the hereditary material, sexual reproductive continuity in space generates geographic populations through life-cycle closure. In turn, biochemical drift within populations reinforced by life-cycle selection (i.e., the selective losses from the appropriate populations) may generate reproductive isolation. With population regulation being considered a spatially limited phenomenon as well as a food-chain one, the mechanism of the origin of reproductive isolation is a focus of the selection process and not an incidental by-product of other adaptations.

In addition, population richness as a key characteristic of species (from one to many populations) is interpreted simply under the member/vagrant hypothesis. One does not have to invoke the existence of different genetic systems for monotypic species, as has been hypothesized by Carson (1975), to account for aspects of population richness. Richness is a function of replication of the physical basis required for life-cycle closure. The absolute abundance of populations and species is a function of the spatial scale of the life cycle that permits continuity and persistence.

The fifth characteristic of species, morphology (both the relative lack of morphological change during some speciation events and the inferred requirement for saltations in form in other events), is also accommodated more easily (even though we have treated the developmental and ontological responses to life-cycle-breaking events as a "black box"). With energetics-type adaptations and speciation considered as separate but contemporaneous processes, one driven by food-chain and the other by life-cycle selection, it is not anomalous that speciation (the development of reproductive isolation and sterility of hybrids) can occur without much morphological change (except for morphological features associated with reproductive isolation itself such as the lock/key structures of *Labidocera* species).

The emphasis on the essential aspect of life-cycle continuity in relation to geographical constraints as the substance of life-cycle selection also allows one to interpret the opposite aspect of morphology in speciation—the inferred existence of saltations in form. Since it is the ontogenies of life cycles that are being selected for, one might expect to observe gradualism in life-cycle changes, but not necessarily gradualism in the morphology of a particular life-history stage. In this view the life cycles, rather than adult morphology, should intergrade gradually. A modest or gradual shift in ontogeny may result in a marked change in morphology at a given developmental stage. The proposed decoupling of energetics adaptation (driven by food-chain selection) from speciation (driven by life-cycle selection) may address the criticism first stated by Mivart (1871), that incipient stages of highly adaptive structures would not be expected to be themselves adaptive (discussed by Allen 1980, p. 368). As long as such incipient stages do not deter "membership" in the population, initially inadaptive features could be carried given that intraspecific competition for resources may be of limited importance.

Finally, duration of species, and thus the frequency of the processes of speciation and extinction, may sit more comfortably under the life-cycle selection rubric. Life-cycle selection is a consevative process that brakes change. The duration of species is a function of the creative forces of biochemical drift and changing physical geography. Without life-cycle-breaking events, and with stability in the biochemical/development system, speciation under life-cycle selection, in spite of changing biological interactions, will not necessarily occur. Population regulation can occur without density dependence in the food-chain or energetics component of the population losses if vagrancy is itself density dependent. Extinction may result at least partially from life-cycle-breaking events generated by spatial changes that cannot be accommodated.

Four additional problems with the evolutionary synthesis were listed (two ecological and two biochemical) which are not characteristics of species themselves.

1. Speciation under life-cycle selection does not emphasize either the key importance of geographical isolation of populations or the importance of small peripheral populations in the evolution of novelties. The processes of speciation do not require these constraints. Reproductive isolation between populations may occur without geographic barriers. The component of population regulation that maintains such isolation generates life-cycle selection. In this respect the hypothesis presented is a retreat to Darwinism. With food-chain selection restricted to energetics adaptation, there is no need to buttress it with geographic isolation to account for speciation.

2. The lack of support for the importance of intraspecific competition for limited resources in population regulation is not an issue under

life-cycle selection, which does not require such competition for population regulation.

3. The "excess" genetic variability discovered since the development of gel electrophoresis and nucleotide sequencing is perhaps less of a problem with life-cycle selection. The masking of inevitable biochemical variability is considered the ultimate function of sexual reproduction and is thus a primary constraint in such selection. The anomaly is not that so much genetic variability exists, but that more does not.

4. The evidence for the molecular clock concept (which in the recent literature has been questioned for life as a whole), or relative constancy in molecular evolution within groups at least, is not inconsistent with life-cycle selection. In spite of the fundamental biochemical constraints, one might expect molecular evolution and speciation to be decoupled processes in time. Species are viewed as conservative entities holding on as long as possible to a particular life-cycle continuity within geographical constraints in spite of the drifting molecular information. To the degree that the molecular information between populations is compatible and development generates the appropriate morphology and behavior for life-cycle continuity, there is cohesiveness in the species even though the imperfect molecular replication of the information system is evolving.

A weakness in the speciation concept is that it does not explore ecological links to the control of development. Ontogeny is defined as a central issue in life-cycle selection; and it is hoped that it has been placed in a more appropriate ecological and geographical framework as a result of the present arguments, but the developmental mechanisms associated with evolving ontogenies are not addressed. Van Valen (1976) captures, in his usual crisp form, the links between development and ecology (p. 180): "Evolution is the control of development by ecology." It is this linkage between ecological processes and changes in ontogenies at the internal level of the organism that has been left out.

···12···

EXPLANATORY POWER OF THE MEMBER/VAGRANT HYPOTHESIS

In this essay, the term *explanatory power* is used in a loose sense. Nothing is actually explained. Rather, a consistency argument is presented whose point of departure is the ultimate function of sexual reproduction itself. The repair hypothesis of Bernstein et al. (1985) on sexual reproduction is a useful framework within which to discuss population regulation in the oceans. The constraint of sexual reproduction involving the "cost of rarity" is central to the member/vagrant hypothesis. If the population regulation hypothesis is robust, such a concept as life-cycle selection emerges from it—in much the same way as the original concept of natural selection was deduced from the assumption of intraspecific competition.

Just as Ghiselin (1969) has stressed that sexual selection can usefully be considered as a separate process from natural selection, it was argued in Chapters 4 and 11 that the broader concept of life-cycle selection (which is defined to include sexual selection) is also usefully considered as a distinct selection process. However, because the natural selection concept itself has evolved to mean temporal changes in the frequency distribution of traits in populations (irrespective of the ecological mechanism generating such changes), the term is not used in its original narrower sense in this essay. To do so might generate unnecessary confusion. Nevertheless, the member/vagrant hypothesis argues that there are two sets of constraints for sexually reproducing populations, one generated ultimately by energy and one by spatial structure. Separate selection processes (food-chain and life-cycle) are defined in relation to the two sets of constraints that generate differential losses from populations.

An advantage of identifying the two ecological components of selection is that speciation and so-called energetics adaptation can be considered as separate processes associated with identifiable components of population regulation. Spatial and energetics constraints are considered to be contemporaneous aspects of population regulation that can usefully be distingushed. The overall set of arguments may provide an alternative perspective on population regulation and speciation in the oceans. From such a perspective a number of contentious issues seem more readily interpretable. It is in this sense, then, that the term "explanatory power" is used.

The member/vagrant hypothesis questions a central concept enunciated by Darwin, that all animal and plant species are striving to increase in numbers and that this ultimately results in competition for limited resources, both inter- and intraspecific. In this essay we conclude that, although it is intuitively logical, the inference that competition for limited resources results from high fecundity in the face of relatively stable abundance on ecological time scales is in fact not necessarily correct. The deduction by Darwin was based on, and is consistent with, the earlier typological species concept that emphasized the adaptive features of adult form. However, from the present understanding of species (involving groups of populations) of sexually reproducing animals with complex life cycles, it is clear that membership itself in a population (i.e., not being a vagrant) is an additional constraint that may be as important as survival. In essence, it does not appear that the "population thinking" of the evolutionary synthesis was carried to its logical conclusion. We would argue that this failure to apply population thinking consistently within the ecological component of evolutionary theory has continued to generate misunderstandings. The evolutionary synthesis did not go far enough in this respect.

Biological Species Concept

The first, and perhaps most fundamental, example of explanatory power of the member/vagrant hypothesis and the life-cycle selection concept is a better understanding or explanation of the biological species concept itself. Quite simply, why are species most frequently, but not universally, groups of populations?

Following the overall set of arguments in this essay, the biological species with its characteristic population pattern and richness is a manifestation of the constraint of the sexual mode of reproduction, which is itself argued to be a constraint of the biochemical instability of the hereditary material. The patterns in populations are generated, not by geographic barriers, but by replicate (or not) physical geographic infrastructure that permits life-cycle closure of populations. Small populations can in fact sit within the distributional limits of larger populations; and, for mobile animals, different populations can share a common distributional area outside the time period of reproduction itself.

The empirical observations on diverse marine fish species suggest that reproductive isolation of populations is generated, not by geographic barriers, but by geographic *opportunities*. Sexual reproduction, because of the "cost of rarity" (Bernstein et al. 1984), requires life-cycle continuity in particular geographic space, and this results in population pattern and thus the "biological species."

Population Richness

Why are some species population rich and others population poor? Both Rensch and Mayr recognized that there are differences between species in this characteristic, but they could not convincingly interpret the patterns. The member/vagrant hypothesis, we believe, plausibly accounts for the patterns in population richness described for marine fish species. Because of the importance of the population concept to fisheries management, pattern and richness of commercially exploited marine species, in the northern Atlantic in particular, are better described than for any other animal group (see Chapter 2). Thus, the fisheries literature may well be the only information base that is sufficient for an analysis of pattern and richness. Richness is simply a function of the existence of replicated geographic infrastructure necessary for life-cycle continuity and is thus physically defined. In this interpretation, observed population patterns are not necessarily evidence of the process of speciation; they may be replicate manifestations of species-specific life cycles in relation to an appropriate geographical framework.

Fecundity

It is of interest, given the key role of fecundity in evolutionary theory (and the observation that fecundity varies dramatically between species), that there is no general theory accounting for the observed pattern. Darwin linked fecundity to the degree of competition that must exist during the life history, but he noted that absolute abundance and fecundity are unrelated. Subsequent developments in ecology have not generalized on species-specific differences in fecundity, nor have they substantiated Darwin's hypothesis, although there is a large and recent literature on the costs of reproduction arising from the life-history strategy school. To our knowledge there has been no demonstration that differences in accumulated life-history competition for limited resources are associated with differences between species in relative fecundity.

The member/vagrant hypothesis generates a simple interpretation of fecundity patterns among species. Absolute fecundity is considered a function of the accumulated dispersive losses associated with the life cycle of the population in relation to the particular physical geography. It is thus a constraint of the spatial scale of the life cycle of the population, and of the physical dispersive characteristics of the population's geographical infrastructure. Long-distance dispersal at various life-history stages is considered an artifact of population persistence within particular geographical constraints rather than the adaptive function of such stages.

We are arguing that fecundity is a function of the accumulated life-cycle "diffusive" losses associated with the particular geographical context underlying populations of species; it is not a function of accumulated life-history intraspecific competition. A general interpretation of species-specific differences in fecundity is the third example of the explanatory power of the population regulation hypothesis.

Absolute Abundance

The controlling mechanism of absolute abundance can perhaps be considered a part of the population richness question. It has been separately identified here because of the importance of "effective" population abundance in population genetics.

Absolute abundance, like fecundity, can be physically defined in relation to the spatial scale of the life cycle. This is best conceptualized by using an example. Cod populations whose early life histories are associated with large "retention" areas such as the Norwegian and Labrador currents are considerably larger than those whose early life histories are associated with small banks such as the Flemish Cap or the Faroe Islands. The juveniles and adults of separate populations of different abundance levels may share a common distributional area. In this view, the physical retention capacity of the early life-history distributional area defines the mean absolute abundance of the population. The location of spawning itself has to be appropriate to generate members rather than vagrants. Animals diffused or advected out of their distributional area may not die, but they may not recruit to the adult population. If vagrancy is at times density dependent, competition for resources is not required for the regulation of absolute abundance of marine populations.

In this sense, absolute abundance may be physically defined in relation to the spatial scale of the geographic framework associated with life-cycle continuity that ensures population persistence. To our knowledge there is no other interpretation in the literature that accounts for the accumulated observations on the mean absolute abundance of marine populations (which range over several orders of magnitude for populations of some fish species). Under life-cycle selection, the specific location of an animal in relation to the appropriate distributional area for the population is as critical as how competitive an individual is. Thus, all features associated with knowing where one is in geographical space are highly selected for. This emphasis on space is a comfortable framework within which to conceptualize the remarkable behavioral features that are associated with, for example, homing.

The debate on the relative importance of density-dependent versus density-independent regulation of population abundance, which was particularly intense during the 1950s and 1960s, dissipated without clear resolution. In general, field studies were not based on observations of self-sustaining populations. Abundance was not usually estimated; instead, density fluctuations were observed at selected study sites. From the perspective of the member/vagrant hypothesis, if vagrancy itself is density dependent at some phase of the life cycle, density dependence in energetics or food-chain processes is not necessarily required for regulation of abundance. Relative stability in numbers of a population can be sustained without density-dependent food limitation or mortality caused by predation and disease. Intraspecific competition for resources, or for avoiding predators, is not a requirement for control.

The main reason that the population regulation issue was not resolved may have been that populations themselves were not usually the unit of investigation. Ecological studies do not often use the biological species concept as a working tool; and in general, the population thinking of the evolutionary synthesis has had little impact on ecology (Kingsland 1985). The roots of the density-dependent/density-independent debate were generated from the studies of pest control. Control of density of insects of a particular *species*, rather than the abundance of a geographically based *population*, was the question of interest.

This emphasis on density of individuals, rather than abundance of individuals in a population, may have hindered the development of population ecology. It may also have prevented the integration of population genetics with ecology. Population genetics actually deals with abundance of populations (i.e., the effective population size is the shared gene pool), whereas population ecology usually deals with density. The lack of explanatory power in population ecology may well be a result of inappropriate units of study in nature. The study of density may be inappropriate to resolve the question of abundance. Population thinking is to a large degree absent.

Parthenogenesis

Parthenogenesis has received considerable attention in the ecological and evolutionary literature. Because of the importance of this literature in arguments dealing with the use of sex to generate increased genetic variability, it is of interest to evaluate pelagic parthenogenesis in the context of the member/vagrant hypothesis (which is an extension of the repair hypothesis for the use of sex). Consideration of the population regulation hypothesis in relation to an elegant contribution by Gerritsen (1980) leads to a different interpretation of the observed distributional pattern of parthenogenesis for zooplankton in the pelagic environment, freshwater and marine. Three observations on aquatic parthenogenetic animals are considered: their requirement for partially bounded physical systems, their size spectrum relative to other zooplankton, and their particular timing of sexual reproduction. The distribution of parthenogenesis, from the perspective of the member/vagrant hypothesis, is explained in relation to the constraint of the "cost of rarity," which is size dependent.

The overall geographical pattern of zooplankton parthenogenesis is summarized by Gerritsen (pp. 736–738). A large percentage of the zooplankton species in small lakes are parthenogenetic animals, whereas in large lakes the percentage is much smaller. Gerritsen estimated the proportion of parthenogenetic species within the total of the observed crustacean species in four sizes of lakes. Fifty-seven percent of the crustacean species were parthenogenetic in lakes of less than 1 hectare, 49% in lakes of 1–100 hectares, 47% in lakes greater than 100 hectares, and 34% in lakes with a water volume greater than 1,000 km^3. Distributional patterns of parthenogenetic species within the larger lakes are also of interest. Cladocerans within the larger lakes are most frequently observed in the partially bounded nearshore areas and at river mouths. For parthenogenetic zooplankton, then, the geographical patterns of distributions are partially a function of bounded physical systems (small lakes, river mouths, estuaries).

Two other distributional features can also be generalized: the size spectrum of parthenogenetic zooplankton species, and the periodicity of sexual reproduction. Parthenogenetic zooplankton (which are generally less than several millimeters in body length) are at the small end of the size spectrum for zooplankton. Gerritsen has demonstrated theoretically that smaller animals require higher population densities to ensure sufficient encounters between individuals of a population for sexual reproduction. His model predicts the minimum population density required (which he defines as the critical density) for successful sexual reproduction in animals. From empirical observations on mode of reproduction (sexual versus parthenogenetic), body length, and critical population density, he suggests (p. 731) that there may be a minimum size for obligate outbreeding zooplankton:

> In North America there are no freshwater copepods smaller than 0.5 mm, and most are larger than 1 mm. . . . Most planktonic rotifers are under 1 mm, mode is 0.2 to 0.4 mm . . . and cladocerans range from 0.2 to 18 mm, with a mode of 1–1.5 mm. . . . Copepods reproduce sexually, while cladocerans and rotifers are usually parthenogenetic. This would suggest that the minimum size for exclusive sexual reproduction among zooplankton is approximately 0.5 mm.

The third feature is the timing of sexual reproduction in cyclic parthenogenetic zooplankton species. Sexual reproduction always occurs at peak population abundance levels. Most interpretations of the timing of the sexual phase stress that it precedes the season of greatest environmental variability and the sexual reproduction mode generates more population variability to exploit or cope with such environmental variability. In these interpretations of parthenogenesis it seems odd that peak population density always precedes the time of maximal environmental variability. In our view, previous interpretations or hypotheses have attempted to explain away the wrong aspect of the switch in sexual modes, and this emphasis has been encouraged by the extant theory of the selective advantages of sexual reproduction (i.e., that it generates genetic variability). We consider the three features of parthenogenetic zooplankton (geographic patterns, size spectrum, and timing of sexual mode) from the reverse perspective: why do the parthenogenetic zooplankton species reduce the frequency of sexual reproduction? not, why do such species have occasional sexual generations at all?

In this view, parthenogenesis is considered a life-history approach that enables a *sexually* reproducing animal to persist in the pelagic environment at the small end of the size spectrum. For animals less than a few millimeters in body length, high population densities are required to permit sufficient encounter between individuals for sexual reproduction. To reach such densities, several generations of asexual reproduction are required. However, in a physically unbounded system, such as the open ocean or central parts of the Great Lakes, the larger scale of turbulent diffusion may physically limit the population density that can be attained even with interspersed asexual generations. In a partially bounded system with particular behavioral characteristics, the dispersal may be less critical.

In this interpretation, the distribution of zooplankton parthenogenesis is considered a function of several constraints:

- the requirement for sexual reproduction at a certain minimal frequency, according to the hypothesis of Dougherty (1955) and Berstein et al. (1984, 1985) that the advantage of sexual reproduction is the conservation of biochemical information;

- the critical population density required for individual sexual encounter in the pelagic environment in relation to animal size; and

- the physical constraint of turbulent diffusion as a function of volume of the water mass.

In sum, cyclical parthenogenesis is a way to ensure sexual reproduction of small animals once in a while for population persistence in a physically constraining environment.

Hull's Paradox

Hull (1984) identifies two aspects of evolutionary theory that are paradoxical. First, the evolutionary synthesis considers sexual reproduction to be maladaptive. Second, Lamarckian inheritance should have been selected for. He states (p. liv): "If only inheritance were Lamarckian, evolution would be rapid, orderly, and efficient. . . . If inheriting acquired characters would be so adaptive, why has a mechanism not evolved to permit it to occur and, once evolved, rapidly become prevalent?" Hull in fact goes a long way toward resolving the paradox that he has defined for the coupled processes of sexual reproduction and replication. In essence he suggests that it may be Lamarckian inheritance, not sexual reproduction, that is maladaptive. It is a provocative discussion.

From the perspective developed here the paradox is nonexistent. Speciation is viewed as an inevitable result of trying to stay the same with an imperfect biochemical mechanism of replication within a nonequilibrium geographical matrix to which life cycles of populations are adapted to ensure persistence. Sexual reproduction under the repair hypothesis of Bernstein et al. (1985) enhances persistence at the biochemical level. Lamarckian inheritance would not be expected to be selected for in this conservative view of the mechanism of speciation. Inheritance through sexual reproduction is treated as a mechanism that enhances conservation of the genome. Inheritance of acquired features would generate the opposite phenomenon, a change in the genome, which would be incompatible with the inferred ultimate function of sexual reproduction.

Collapse of Life Cycles

We mentioned earlier that one trend in evolution has been the development of complex life cycles, as well as the reverse process of the collapse, or internalization, of complex life cycles. Embryos in this view are internalized free-living stages. As already argued, the evolution of sexual reproduction created order out of chaos, in a distributional sense at least. Subsequent to the evolution of sexual reproduction, the associated requirement for union of separate individuals during or at the end of the life cycle generates populations and persistent patterns in relation to geographical constraints.

Two trends can be identified resulting from the constraint of sexual reproduction and the dispersive features of physical geography. On the one hand, complex life cycles have evolved to ensure sufficient encounter between individuals and thus population persistence. In other words, a variety of complex solutions have evolved which "use" the physical geography (as well as other species) for life-cycle continuity. On the other hand, complex life cycles have tended toward internalization to become independent of the constraints of the geography (i.e., to eliminate stages influenced by diffusion). This trend is not

pronounced in the pelagic environment but is well developed at interfaces between the water and the sea floor and between air and land.

It is argued that life-cycle selection involving life-cycle-breaking events caused by changing physical geography has played a major
role in this back-and-forth process of the collapse and expansion of
complex life cycles. In the oceans (except perhaps in coral reefs and
kelp forests) the changing physical environment itself provides the
infrastructure for possible solutions to life-cycle continuity. In terrestrial systems the structure of the living components, the rich mosaic
of the ecosystem (both biotic and geographical) as a whole, is available
for various solutions to life-cycle continuity. Needham (1930) discusses
the observation that many animal groups have not invaded freshwater.
Also, many lake species do not have a planktonic phase of the life cycle.
The flushing characteristics of rivers may inhibit the invasion of forms
with complex life histories involving a planktonic phase. Life-cycle
closure may not be feasible in the river system for many forms. If so, a
"collapsed" life cycle would favor invasion from the oceans to lakes.
Some of the anomalous differences observed between lakes and the
oceans may be a function of physical geographic constraints on life
cycles within *rivers* (which link the oceans to the lakes).

The trend in the evolution of the collapse of complex life cycles,
involving the internalization of early life-history stages, reduces the
importance of spatial processes to population regulation. In the pelagic
environment, however, this trend has not been pronounced; complex
life histories prevail. Thus, under the member/vagrant hypothesis,
spatial processes should predominate in population regulation. At
interfaces the internalization is more often observed. One would expect that, as the dependence on spatial processes for life-cycle closure
is minimized, food-chain processes would be more important in population regulation. In parallel, evolution may be associated more with
morphological changes reflecting food-chain specialization. A comparison of the results of evolution of amphipods (which have a partially
collapsed life cycle) to copepods (which do not) is supportive. Relative
freedom from the constraints of free-crossing by the collapse of life
cycles should be associated with a greater role of food-chain processes
in the regulation of abundance and thus a larger role for energetics
adaptations in evolution. Amphipod species diversity (and the nature of
the morphological adaptations) may be consistent with this reasoning.
Pelagic copepod speciation, however, frequently has involved no ener-

getics-related adaptations. Not to put too fine a point on it at this stage, the roles of energy and spatial structure in population regulation should vary according to the degree of collapse in the life cycle. This reasoning may provide a simple test of the member/vagrant hypothesis (i.e., an analysis of the morphological differences between species of genera that have complex and collapsed life cycles).

Concluding Remarks

The initial focus of this essay was limited to population regulation in the oceans. However, since it is generalized that spatial processes predominate in marine population regulation and that intraspecific competition for limited resources or predator avoidance is less important to this particular process, the implications of this generalization at the ecological level on evolutionary theory are evaluated. If in fact predation and intraspecific competition for limited resources are not necessarily critical to marine population regulation, natural selection as narrowly defined in Darwinism must play a lesser role in evolution than is now hypothesized. Further, a historical sketch of evolutionary theory indicates that the ecological aspects of the origin of reproductive isolation (i.e., of speciation) are not well understood. The concluding part of this essay explored the role of ecological processes in speciation and adaptation, arguing that life-cycle selection is important in speciation in the oceans, with food-chain selection important to energetics adaptations. Speciation and the evolution of this class of adaptation are considered to be contemporaneous processes that are associated with different ecological phenomena (i.e., different components of population regulation). Spatial structure and food-chain events generate separate sets of ecological constraints. We address briefly here how this modification of the components of natural selection fits in with neo-Darwinism and with the evolutionary synthesis.

The essential elements of neo-Darwinism, a term which was coined to encompass Weismann's elaboration and defense of Darwinism, are that (1) the hereditary material is "hard" rather than "soft" (i.e., it is not modified by acquired characteristics) and (2) natural selection is a sufficient mechanism for both adaptation and speciation. It was an elaboration in that a new theory of heredity was added (the germ theory), yet a narrowing of Darwinism by a heavier, perhaps exclusive, emphasis on natural selection as a sufficient mechanism for all aspects of evolution.

The evolutionary synthesis is a much more extensive development than neo-Darwinism that at present, because of its shifting dimensions, is difficult to define precisely. The ecological and natural history contributions to the synthesis, however, are easier to identify: (1) the biological species concept itself (a shared contribution with the systematists); (2) an emphasis on geographic isolation of populations in evolution; (3) a reduced role for natural selection in speciation; and (4) a continued emphasis on competition for limited resources as a core component of natural selection.

The present treatment covers (1) the acceptance in the repair hypothesis of the sexual mode of reproduction (Bernstein et al. 1985); (2) the member/vagrant hypothesis of population regulation; and (3) the life-cycle selection concept of speciation. It continues certain trends that are discernible in the synthesis. The importance of the biological species concept is reinforced, and a narrowly defined concept of food-chain selection is argued to play a reduced role (limited to energetics adaptations). What is different is the questioning of the necessity of geographic barriers (in particular the heavy emphasis of genetic revolutions in peripherally isolated populations) in speciation and the evolution of novelties, and a reduced role for competition for limited resources in this process.

Although the shifts in emphasis are modest, because they are at the conceptual core of the mechanism of evolution (i.e., population regulation), the implications may be substantive. For example, population richness is not necessarily evidence of speciation in action. Particular location in space in relation to continued membership in a population, rather than increases in numbers, is considered central to the selection process for sexually reproducing animals. In sum, the arguments in this essay dealing with population regulation and speciation in the oceans are consistent with much of the synthesis (i.e., emphasis on the individual in the selection process, acceptance of hard inheritance), but they continue the trend of reducing the role of competition for resources in the speciation process itself.

The arguments developed here may contribute to the present debate on the importance of considering hierarchies in evolution. The "paradox of the first tier" (Gould 1985) may be resolved without having to consider higher levels than the first tier. The paradox defined by Gould is the overall lack of evidence for the progress in the "tree of life" that is predicted *if* competition is the mechanism of evolution. By progress Gould means not that adaptations are not progressive, nor that there is not evidence for increased complexity in life through evolutionary time. Rather, he means that the linkages through time in the genealogy of life forms have not been progressive. Increases in complexity, which we may define as progress, come from haphazard sources. This, then, is the defined paradox—lack of evidence for progress in the *sources and linkages* of increases in complexity coupled with a theory of evolution that infers such a progression.

Gould resolves the paradox by arguing that processes at higher levels (species selection and massive extinctions due to abiotic events) are responsible for reordering the results of processes at the first tier (i.e., speciation and adaptation by natural selection). We argue that from the viewpoint presented in this essay there may be no paradox, since modest changes in the conceptualization of evolutionary processes at the first tier do not predict progress in the sense inferred by Gould. Speciation is viewed here as the result of attempting to stay the same in the face of both internally and externally generated changes at, respectively, the molecular level (which disrupts the ability of the hereditary material to reproduce the developmental program) and the spatial dimension (which disrupts the ability of the population to complete the necessary life-cycle continuity to ensure its persistence). Speciation in sexually reproducing animals is considered a conservative process in nonequilibrium internal (to the individual) and external (to the population) environments. In this sense, speciation by life-cycle selection does not generate "progress."

To the degree that this restatement of the processes at the first tier eliminates the paradox, the lack of progress (as specifically defined by Gould) does not have to be explained at higher levels. In addition, the sexual mode of reproduction itself is not a characteristic of species that has to be explained by species selection at the second tier. Taking competition for limited resources out of the process of speciation but leaving it in the process generating energetics adaptations may resolve the paradox of "progressive" adaptation in evolution without evidence for "progress" in the linkages.

On an ecological time scale, the generalization of the member/vagrant hypothesis dealing with marine populations may provide an explanation for some difficulties that have been identified in the integration of population genetics and theoretical ecology. Ayala (1983) briefly states the nature of the problem. With respect to population genetics, he states that theoretical developments early in this century preceded empirical observations. The theoretical developments in ecology, in contrast, developed later, in the 1960s, much of it emanating from the Hutchinson and MacArthur school, which emphasizes the critical role of competition for resources in ecology. Ayala concludes that the accomplishments of the accumulated research in the two disciplines have not been satisfactorily interpreted to explain the evolution of natural populations.

From the perspective of the arguments developed in this essay this shortfall in accomplishments may well be a flaw in the theoretical developments in ecology. Competition for limited resources is *assumed* to be of major importance in the regulation of abundance. The empirical observations on animal populations in the oceans do not in our reading of the literature justify this assumption. A large component of the excess fecundity is interpreted to be a function of spatial constraints to membership in populations. Theoretical ecology in a certain sense put aside the controversy on population regulation in the 1960s and accepted intuitively that intraspecific competition for resources is important.

The inferred lack of validity of this assumption, as well as the lack of understanding of the geographic basis of populations, may have contributed to the slow progress in comprehending, in Ayala's words, the evolution of natural populations. Population ecology, including both field studies and theoretical work, did not build upon the population systematics literature that generated the *Rassenkreis* and the biological species concept. Population genetics, in contrast, has strong links to population systematics, thanks to the bridging role played by Chetverikov and Dobzhansky in particular. Population genetics requires an understanding of effective population size. Population ecology rarely addresses this concept.

The marine fisheries literature, however, because of the importance of population thinking to fisheries management (starting with Hjort in 1914), is firmly rooted upon the initial contributions of Heincke on population systematics of Atlantic herring. Between 1880 and 1930, the golden age of fisheries research, substantive increases in understanding population ecology were generated by the marine research community in northern Europe (as represented by the contributions of Damas, Fulton, Heincke, Hjort, and Schmidt). In this essay we have addressed the same questions as posed by these earlier fisheries biologists, with the added perspective of the more recent literature on marine biology and oceanography. The accumulated empirical observations on marine fish populations may provide insights into the underlying causes of pattern and richness, which could contribute to a better understanding of the evolution of natural populations.

ENDNOTES

Page 17 [1]"Nous montrerons combien l'aire de reproduction est énor-
mement plus restreinte que l'aire totale de dispersion, que la
région où la ponte s'effectue principalement est différente
pour les diverse espèces, qu'elle est définie par des conditions hydrographiques."

Page 17 [2]"L'idée fondamentale a été de suivre les germes (oeufs, larves
et jeunes alevins) pendant leur dispersion progressive sous
l'influence des courants, leur migration passive depuis les
aires de ponte jusqu'aux limites géographiques de l'espèce."

Page 45 [3]"Anfang Januar wird die Masse der im November geschlüpf-
ten Larven vor der Küste der holländischen Provinz Zeeland
angetroffen. Einige Beobachtungen sprechen dafür, dass
solche hier in einem kleinen Stromwirbel vor der Rhein-
mündung noch längere Zeit zurückbleiben. Im ganzen aber
schreitet die nordostwärts gerichtete Verfrachtung fort."

Page 118 [4]"L'espèce se maintient grâce à l'existence dans ces régions
d'un courant circulatoire que ramène périodiquement un
certain proportion des individus à la surface de l'océan et
entrainés dans le mouvement continuel des eaux. . . . Le
mécanisme de la circulation joue donc ici le rôle principal
pour la conservation de l'espèce."

Literature Cited

Aldredge, A. L., B. H. Robinson, A. Fleminger, J. J. Torres, J. M. King, and W. H. Hamner. 1984. Direct sampling and in situ observation of a persistent copepod aggregation in the mesopelagic zone of the Santa Barbara Basin. *Marine Biology* 80:75–81.

Allen, G. E. 1980. The evolutionary synthesis: Morgan and natural selection revisited. In *The Evolutionary Synthesis,* ed. E. Mayr and W. B. Provine (Harvard University Press, Cambridge), pp. 356–382.

Allen, J. A. 1959. On the biology of *Pandalus borealis* Kroyer with reference to a population off the Northumberland coast. *Journal of the Marine Biological Association U. K.* 38:189–220.

————. 1966. The dynamics and interrelationships of mixed populations of Caridea found off the north-east coast of England. In *Some Contemporary Studies in Marine Science,* ed. H. Barnes (George Allen & Unwin, London), pp. 45–66.

Anderson, D. M. and K. D. Stolzenbach. 1985. Selective retention of two dinoflagellates in a well-mixed estuarine embayment: The importance of diel vertical migration and surface avoidance. *Marine Ecology Progress Series* 25:39–50.

Anderson, J. T. 1982. *Distribution, abundance, and growth of cod* (Gadus morhua) *and redfish* (Sebastes *spp.*) *larvae on Flemish Cap, 1981.* Northwest Atlantic Fisheries Organization, SCR Document 82/VI/37.

Auger, F. and H. Powles. 1980. *Estimation of the herring spawning biomass near Isle Verte in the St. Lawrence Estuary from an intensive larval survey in 1979.* Canadian Atlantic Fisheries Scientific Advisory Committee, Research Document 80/59.

Ayala, F. J. 1983. Foreword. In *Theory of Natural Selection and Population Growth*, ed. L. R. Ginzburg (Benjamin/Cummings, Menlo Park), pp. vii–ix.

Bahr, L. M., Jr. 1982. Functional taxonomy: An immodest proposal. *Ecological Modelling* 15:211–233.

Bailey, K. M. 1981. Larval transport and recruitment of Pacific hake *Merluccius productus*. *Marine Ecology Progress Series* 6:1–9.

Baker, R. R. 1978. *The Evolutionary Ecology of Animal Migration.* Holmes and Meier, New York, 1012 pp.

Barlow, J. P. 1955. Physical and biological processes determining the distribution of zooplankton in a tidal estuary. *Biological Bulletin* (Woods Hole) 109:211–225.

Barnett, A. M., A. E. Jahn, P. D. Sertic, and W. Watson. 1984. Distribution of ichthyoplankton off San Onofre, California, and methods for sampling very shallow coastal waters. *Fisheries Bulletin* (U.S.) 82:97–111.

Bateson, W. 1909. Heredity and variation in modern lights. In *Darwin and Modern Science*, ed. A. C. Seward (Cambridge University Press, London), pp. 85–101.

Bennett, J. H. 1983. *Natural Selection, Heredity, and Eugenics.* Clarendon Press, Oxford. 306 pp.

Berger, W. H. and J. S. Killingley. 1982. The Worthington effect and the origin of the younger Dryas. *Journal of Marine Research* 40(Suppl.):27–38.

Berger, W. H., R. C. Finkel, J. S. Killingley, and V. Marchig. 1983. Glacial-Holocene transition in deep-sea sediments: Manganese-spike in the east-equatorial Pacific. *Nature* 303:231–233.

Berger, W. H., J. S. Killingley, and E. Vincent. 1985. Timing of deglaciation from an oxygen isotope curve for Atlantic deep-sea sediments. *Nature* 314:156–158.

Berggren, W. A. and C. D. Hollister. 1977. Plate tectonics and paleo-circulation—commotion in the ocean. *Tectonophysics* 38:11–48.

Berkes, F. 1976. Ecology of euphausiids in the Gulf of St. Lawrence. *Journal of the Fisheries Research Board of Canada* 33:1894–1905.

Bernstein, H., H. C. Byerly, F. A. Hopf, and R. E. Michod. 1984. Origin of sex. *Journal of Theoretical Biology* 110:323–351.

————. 1985. Genetic damage, mutation, and the evolution of sex. *Science* 229:1277–1281.

Berrien, P. L., 1978. Eggs and larvae of *Scomber scombrus* and *Scomber japonicus* in continental shelf waters between Massachusetts and Florida. *Fisheries Bulletin* (U.S.) 76:95–115.

Beverton, R. J. H. and S. J. Holt. 1957. *On the Dynamics of Exploited Fish Populations.* U.K. Ministry of Agriculture, Fisheries and Food, Fishery Investigations, ser. 2, vol. 19. 533 pp.

Binet, D. and E. Suisse de Sainte Claire. 1975. Le copépode planctonique *Calanoides carinatus:* Répartition et cycle biologique au large de la côte d'Ivoire. *Cahiers O.R.S.T.O.M.*, Série Océanographie 13:15–30.

Birch, L. C. 1970. The role of environmental heterogeneity and genetical heterogeneity in determining distribution and abundance. In *Dynamics of Population*, ed. P. J. den Boer and G. R. Gradwell (Centre for Agricultural Publication and Documentation, Wageningen, Netherlands), pp. 109–129.

Blaxter, J. H. S. 1985. The herring: A successful species? *Canadian Journal of Fisheries and Aquatic Sciences* 42(Suppl. 1):21–30.

Boden, B. P. 1952. Natural conservation of insular plankton. *Nature* 169:697–699.

Boden, B. P. and E. M. Kampa. 1953. Winter cascading from an oceanic island and its biological implications. *Nature* 171:426–427.

Boëtius, J. and E. F. Harding. 1985. A re-examination of Johannes Schmidt's Atlantic eel investigations. *Dana* 4:129–162.

Bolz, G. R. and R. G. Lough. 1984. Retention of ichthyoplankton in the Georges Bank region the autumn–winter seasons, 1971–77. *Journal of Northwest Atlantic Fishery Science* 5:33–45.

Bonner, J. T. 1965. *Size and Cycle.* Princeton University Press, Princeton. 219 pp.

Bosch, H. F. and W. R. Taylor. 1970. *Ecology of* Podon polyphemoides *(Crustacea, Branchiopoda) in the Chesapeake Bay.* Chesapeake Bay Institute, Ref. 70-5. 77 pp.

————. 1973. Distributions of the cladoceran *Podon polyphemoides* in the Chesapeake Bay. *Marine Biology* 19:161–171.

Boucher, J. 1982. Peuplement de copépodes des upwellings côtiers Nord-Ouest africains. II. Maintien de la localisation spatiale. *Oceanologica Acta* 5:199–207.

————. 1984. Localization of zooplankton populations in the Ligurian Sea front: Role of ontogenic migration. *Deep-Sea Research* 29:953–965.

Boucher, J., F. Ibanez, and L. Prieur. 1987. Daily and seasonal variations in the spatial distribution of zooplankton populations in relation to the physical structure in the Ligurian sea front. *Journal of Marine Research* 45:133–173.

Bousfield, E. L. 1955. *Ecological control of the occurrence of barnacles in the Mirimichi Estuary.* National Museum of Canada, Bulletin 137. 89 pp.

Bowers, A. B. 1952. Studies on the herring *(Clupea harengus L.)* in Manx waters: The autumn spawning, and the larval and post-larval stages. *Proceedings of the Liverpool Biological Society* 58:47–74.

Boyar, H. C., R. R. Marak, F. E. Perkins, and R. A. Clifford. 1973. Seasonal distribution and growth of larval herring *(Clupea harengus L.)* in the Georges Bank–Gulf of Maine area from 1962 to 1970. *Journal du Conseil* 35:36–51.

Brander, K. M. and R. R. Dickson. 1984. An investigation of the low level of fish production in the Irish Sea. *Rapports et Procès-verbaux des Réunions, Conseil international pour l'Exploration de la Mer* 183:234–242.

Bretsky, P. W. and S. M. Klofak. 1985. Margin to craton expansion of late Ordovician benthic marine invertebrates. *Science* 227:1469–1471.

Bridger, J. P. 1961. On fecundity and larval abundance of Downs herrings. *Fishery Investigations II* (G.B.) 13(3):1–30.

Brinton, E. 1962. The distribution of Pacific euphausiids. *Bulletin of the Scripps Institution of Oceanography* 8:51–270.

Brown, J. H. and A. C. Gibson. 1983. *Biogeography.* C. V. Mosby, St. Louis. 643 pp.

Bückmann, A. 1950. Die Untersuchungen der biologischen Anstalt über die Ökologie der Heringsbrut in der südlichen Nordsee, I. *Helgoländer Wissenschaftliche Meeresuntersuchungen* 3:1–57.

Burton, R. S. and M. W. Feldman. 1982. Population genetics of coastal and estuarine invertebrates: Does larval behavior influence population structure. In *Estuarine Comparisons*, ed. V. Kennedy (Academic Press, New York), pp. 537–551.

Buzas, M. A. and S. J. Culver. 1984. Species duration and evolution:

Benthic foraminifera on the Atlantic continental margin of North America. *Science* 225:829–830.

Caffey, H. M. 1985. Spatial and temporal variation in settlement and recruitment of intertidal barnacles. *Ecological Monographs* 55:313–332.

Campbell, A. and R. K. Mohn. 1983. Definition of American lobster stocks for Canadian maritimes by analysis of fishery-landing trends. *Transactions of the American Fisheries Society* 112:744–759.

Carriker, M. R. 1951. Ecological observations on the distributions of oyster larvae in New Jersey estuaries. *Ecological Monographs* 21:19–38.

Carruthers, J. N., A. L. Lawford, V. F. C. Veley, and B. B. Parrish. 1951. Variations in brood-strength in North Sea haddock, in the light of relevant wind conditions. *Nature* (London) 168:317–319.

Carscadden, J. E. and W. C. Leggett. 1975. Meristic differences in spawning populations of American shad, *Alosa sapidissima:* Evidence for homing to tributaries in the St. John River, New Brunswick. *Journal of the Fisheries Research Board of Canada* 32:653–650.

Carson, H. L. 1975. The genetics of speciation at the diploid level. *American Naturalist* 109:83–92.

Chelton, D. B., P. A. Bernal, and J. A. McGowan. 1982. Large-scale interannual physical and biological interaction in the California Current. *Journal of Marine Research* 40:1095–1125.

Chetverikov, S. S. 1926. On certain aspects of the evolutionary process from the standpoint of modern genetics [in Russian]. *Zhurnal Eksperimental'noi Biologii* A2:3–54. Trans. M. Barker, ed. I. M. Lerner, *Proceedings of the American Philosophical Society* 105(2):167–195(1961).

Clark, R. S. 1933. Herring larvae: The mixing of the broods in Scottish waters. *Rapports et Procès-verbaux des Réunions, Conseil Permanent international pour l'Exploration de la Mer* 85(3):11–18.

Cole, L. C. 1954. The population consequences of life history phenomena. *Quarterly Review of Biology* 27:103–137.

———. 1957. Sketches of general and comparative demography. *Cold Spring Harbor Symposia on Quantitative Biology* 22:1–15.

Colebrook, J. M. 1978. Continuous plankton records: Zooplankton and environment, north-east Atlantic and North Sea, 1948–1975. *Oceanologica Acta* 1:9–23.

———. 1979. Continuous plankton records: Seasonal cycles of phytoplankton and copepods in the North Atlantic Ocean and the North Sea. *Marine Biology* 51:23–32.

———. 1981. Continuous plankton records: Persistence in time-series of annual means of abundance of zooplankton. *Marine Biology* 61:143–149.

———. 1982a. Continuous plankton records: Phytoplankton, zooplankton, and environment, north-east Atlantic and North Sea, 1958–1980. *Oceanologica Acta* 5:473–480.

———. 1982b. Continuous plankton records: Persistence in time-series and the population dynamics of *Pseudocalanus elongatus* and *Acartia clausi*. *Marine Biology* 66:289–294.

———. 1982c. Continuous plankton records: Seasonal variations in the distribution and abundance of plankton in the north Atlantic Ocean and the North Sea. *Journal of Plankton Research* 4:435–462.

———. 1984. Continuous plankton records: Relationships between species of phytoplankton and zooplankton in the seasonal cycle. *Marine Biology* 83:313–323.

———. 1985. Continuous plankton records: Overwintering and annual fluctuations in the abundance of zooplankton. *Marine Biology* 84:261–265.

Colebrook, J. M. and A. H. Taylor. 1984. Significant time scales of long-term variability in the plankton and the environment. *Rapports et Procès-verbaux des Réunions, Conseil international pour l'Exploration de la Mer* 183:20–26.

Colton, J. B., Jr. and R. F. Temple. 1961. The enigma of Georges Bank spawning. *Limnology and Oceanography* 6:280–291.

Connell, J. H. 1961. The effects of competition, predation by *Thais lapillus*, and other factors on natural populations of the barnacle *Balanus balanoides*. *Ecological Monographs* 31:61–104.

Coope, G. R. 1979. Late Cenozoic fossil coleoptera: Evolution, biogeography, and ecology. *Annual Review of Ecology and Systematics* 10:246–267.

Cronin, T. M. 1985. Speciation and stasis in marine ostracoda: Climatic modulation of evolution. *Science* 227:60–63.

Cronin, T. W. 1982. Estuarine retention of larvae of the crab *Rhithropanopeus harrisii*. *Estuarine Coastal Shelf Science* 15:207–220.

Cronin, T. W. and R. B. Forward, Jr. 1982. Tidally timed behavior: Effects on larval distributions in estuaries. In *Estuarine Comparisons*, ed. V. Kennedy (Academic Press, New York), pp. 505–520.

Cushing, D. H. 1961. On the failure of the Plymouth herring fishery. *Journal of the Marine Biological Association U. K.* 41:719–816.

———. 1975. *Marine Ecology and Fisheries*. Cambridge University Press, London. 278 pp.

———. 1986. The migration of larval and juvenile fish from spawning ground to nursery ground. *Journal du Conseil* 43:43–49.

Cushing, D. H. and J. G. K. Harris. 1973. Stock and recruitment and the problem of density dependence. *Rapports et Procès-verbaux des Réunions, Conseil international pour l'Exploration de la Mer* 104:142–155.

Dadswell, M. J., G. D. Melvin, and P. J. Williams. 1983. Effect of turbidity on the temporal and spatial utilization of the inner Bay of Fundy by American shad *(Alosa sapidissima)* (Pisces:Clupeidae) and its relationship to local fisheries. *Canadian Journal of Fisheries and Aquatic Sciences* 40(Suppl. 1):322–330.

Dahl, K. 1907. The scales of herring as a means of determining age, growth and migration. *Report of the Norwegian Fisheries and Marine Investigations* 2(6):1–36.

Damas, D. 1905. *Notes biologiques sur les copépodes de la Mer Norvégienne*. Conseil Permanent International pour l'Exploration de la Mer, Publication de Circonstance 22. 23 pp.

———. 1909. Contribution à la biologie des gadids. *Rapports et Procès-verbaux des Réunions, Conseil international pour l'Exploration de la Mer* 10(Part 3):1–277.

Damas, D. and E. Koefoed. 1907. Le plancton de la Mer du Grönland. In *Croisière océanographique: Accomplie à bord de la Belgica dans la Mer du Grönland*, comp. Louis Philippe Robert, duc d'Orléans (C. Bulens, Brussels), pp. 347–453.

Darwin, C. 1859. *The Origin of Species*. 1st ed. Penguin English Library, 1982. 476 pp.

David, P. M. 1955. The distribution of *Sagitta gazellae* Ritter-Záhony. *Discovery Reports* 27:235–278.

Davidson, K., J. C. Roff, and R. W. Elner. 1985. Morphological elec-
trophoretic and fecundity characteristics of Atlantic snow crab,
Chionoecetes opilio, and implications for fisheries management.
Canadian Journal of Fisheries and Aquatic Sciences 42:474–482.

Davis, C. S. 1984a. Interaction of a copepod population with the mean
circulation on Georges Bank. *Journal of Marine Research*
42:573–590.

———. 1984b. Food concentrations on Georges Bank: Non-limiting
effect on development and survival of laboratory reared
Pseudocalanus sp. and *Paracalanus parvus* (Copepoda: Calanoida). *Marine Biology* 82:41–46.

———. 1984c. Predatory control of copepod seasonal cycles on
Georges Bank. *Marine Biology* 82:31–40.

de Beer, G. R. 1940. *Embryos and Ancestors*. Clarendon Press, Oxford. 108 pp.

de Lafontaine, Y., M. Sinclair, M. I. El-Sabh, C. Lassus, and R.
Fournier. 1984a. Temporal occurrence of ichthyoplankton in relation to hydrographic and biological variables at a fixed station in
the St. Lawrence Estuary. *Estuarine Coastal Shelf Sciences*
18:177–190.

de Lafontaine, Y., M. I. El-Sabh, M. Sinclair, S. H. Messieh, and J.-D.
Lambert. 1984b. Structure océanographique et distribution spa-
tio-temporelle d'oeufs et de larves de poissons dans l'estuaire
maritime et la partie ouest du Golfe Saint-Laurent. *Sciences et
Techniques de l'Eau* 17:43–50.

de Wolf, P. 1973. Ecological observations on the mechanisms of dispersal of barnacle larvae during planktonic life and settling.
Netherlands Journal of Sea Research 6:1–129.

———. 1974. On the retention of marine larvae in estuaries. *Thalassia
Jugoslavica* 10:415–424.

DeVries, T. J. and W. G. Pearcy. 1982. Fish debris in sediments of the
upwelling off central Peru: A late quaternary record. *Deep-Sea
Research* 28:87–109.

Dittel, A. I. and C. E. Epifanio. 1982. Seasonal abundance and vertical
distribution of crab larvae in Delaware Bay. *Estuaries* 5:197–202.

Dobzhansky, Th. 1937. *Genetics and the Origin of Species*. Columbia
University Press, New York. 364 pp.

Doherty, P. J. 1983. Tropical territorial damselfishes: Is density limited
by aggression or recruitment? *Ecology* 64:176–190.

Dooley, H. D. and D. W. McKay. 1979. The drift of herring larvae from the west coast to the North Sea. *Scottish Fisheries Bulletin* 45:10–12.

Dougherty, E. C. 1955. Comparative evolution and the origin of sexuality. *Systematic Zoology* 4:145–169.

Dover, G. A. 1985. Shadow boxing with Darwin. *Nature* 318:19–20.

Ebert, T. A. 1968. Growth rates of the sea urchin, *Strongylocentrotus purpuratus*, related to food availability and spine abrasion. *Ecology* 49:1075–1091.

————. 1982. Longevity, life history, and relative body wall size in sea urchins. *Ecological Monographs* 55:712–729.

Efford, I. E. 1970. Recruitment to sedentary marine populations as exemplified by the sand crab, *Emerita analoga* (Decapoda, Hippidea). *Crustaceana* 18:293–308.

Ehrlich, P. R. and L. C. Birch. 1967. The "balance of nature" and population controls. *American Naturalist* 101:97–107.

Ehrlich, P. R. and D. D. Murphy. 1981. The population biology of checkerspot butterflies *(Euphydras)*. *Biologisches Zentralblatt* 100:613–629.

Eldredge, N. 1985. *Unfinished Synthesis: Biological Hierarchies and Modern Evolutionary Thought*. Oxford University Press, Oxford. 237 pp.

Ellertsen, B. P., P. Fossum, P. Solemdal, S. Sundby, and S. Tilseth. 1986. *The effect of biological and physical factors on the survival of arcto-Norwegian cod and the influence on recruitment variability*. Northwest Atlantic Fisheries Organization, SCR Doc. 86/116.

Emery, A. 1972. Eddy formation from an oceanic island: Ecological effects. *Caribbean Journal of Science* 12:121–128.

Emlen, J. T. 1986. Land-bird densities in matched habitats on six Hawaiian islands: A test of resource-regulation theory. *American Naturalist* 127:125–139.

Enright, J. T. 1977. Diurnal vertical migration: Adaptive significance and timing. 1. Selective advantage: A metabolic model. *Limnology and Oceanography* 22:856–872.

Enright, J. T. and H.-W. Honegger. 1977. Diurnal vertical migration: Adaptive significance and timing. 2. Test of the model: Details of timing. *Limnology and Oceanography* 22:873–886.

Epifanio, C. E. and A. I. Dittel. 1982. Comparison of dispersal of crab larvae in Delaware Bay, U.S.A., and the Gulf of Nicoya, Central America. In *Estuarine Comparisons*, ed. V. Kennedy (Academic Press, New York), pp. 477–487.

Evans, G. T. 1978. Biological effects of vertical horizontal interactions. In *Spatial Pattern in Plankton Communities*, ed. J. H. Steele (Plenum Press, New York), pp. 157–179.

Findley, J. S. and M. J. Findley. 1985. A search for pattern in butterfly fish communities. *American Naturalist* 126:800–816.

Fisher, R. A. 1958a. *The Genetical Theory of Natural Selection*. 2d ed. Dover, New York. 291 pp.

———. 1958b. Polymorphism and natural selection. *Journal of Ecology* 46:289–293.

Fleminger, A. 1967. Taxonomy, distribution, and polymorphism in the *Labidocera jollae* group with remarks on evolution within the group (Copepoda:Calanoida). *Proceedings of the U.S. National Museum* 120:1–61.

———. 1973. Pattern, number, variability, and taxonomic significance of integumental organs (sensilla and glandular pores) in the genus *Eucalanus* (Copepoda, Calanoida). *Fisheries Bulletin* (U.S.) 71:965–1010.

———. 1975. Geographical distribution and morphological divergence in American coastal zone planktonic copepods of the genus *Labidocera*. In *Estuarine Research*, ed. L. E. Cronin (Academic Press, New York), pp. 392–419.

———. 1985. Dimorphism and possible sex change in copepods of the family Calanidae. *Marine Biology* 88:273–294.

Fleminger, A. and K. Hulsemann. 1973. Relationship of 14 Indian Ocean epiplanktonic calanoids to the world oceans. In *The Biology of the Indian Ocean*, ed. P. Zeitzschel and S. A. Gerlach (Springer-Verlag, Berlin), pp. 339–348.

———. 1974. Systematics and distributions of the four sibling species comprising the genus *Pontellina* Dana (Copepoda, Calanoida). *Fisheries Bulletin* (U.S.) 72:63–120.

———. 1977. Geographical range and taxonomic divergences in north Atlantic *Calanus* (*C. helgolandicus*, *C. finmarchicus* and *C. glacialis*). *Marine Biology* 40:233–248.

Fleminger, A. and E. Tan. 1966. The *Labidocera mirabilus* species group (Copepoda:Calanoida) with a description of a new Bahamian species. *Crustaceana* 11:291–301.

Fortier, L. and W. C. Leggett. 1982. Fickian transport and the dispersal of fish larvae. *Canadian Journal of Fisheries and Aquatic Sciences* 39:1150–1163.

Foxton, P. 1964. Seasonal variations in the plankton of Atlantic waters. In *Biologie Antarctique,* Proceedings of SCOR Symposium, Paris (Hermann, Paris), pp. 2–8.

Frèchet, A., J. J. Dodson, and H. Powles. 1983. Use of variation in biological characteristics for the classification of anadromous rainbow smelt *(Osmerus mordax)* groups. *Canadian Journal of Fisheries and Aquatic Sciences* 40:718–727.

Frost, B. and A. Fleminger. 1968. A revision of the genus *Claus ocalanus* (Copepoda:Calanoida) with remarks on distributional patterns in diagnostic characters. *Bulletin of the Scripps Institution of Oceanography* 12:1–99.

Frost, B. W., M. R. Landry, and R. P. Hassett. 1983. Feeding behavior of large calanoid copepods *(Neocalanus cristatus* and *N. plumchrus)* from the subarctic Pacific Ocean. *Deep-Sea Research* 30:1–13.

Fulton, T. W. 1889. The spawning and spawning places of marine food-fishes. *Fishery Board of Scotland Annual Reports* 8:257–282.

———. 1895. The relation of marine currents to offshore spawning areas and inshore nurseries. *Fishery Board of Scotland Annual Reports* 13:153–164.

———. 1900. Additional notes on the surface currents of the North Sea. *Fishery Board of Scotland Annual Reports* 18:370–381.

Gagné, J. A. and R. N. O'Boyle. 1984. The timing of cod spawning on the Scotian Shelf. In *The Propagation of Cod* Gadus morhua *L.,* ed. E. Dahl, D. S. Danielssen, E. Moksness, and P. Solemdal, Flodevigen Rapportser., vol. 1, pp. 501–517.

Garrett, C. J. R., J. R. Keely, and D. A. Greenberg. 1978. Tidal mixing versus thermal stratification in the Bay of Fundy and Gulf of Maine. *Atmosphere-Ocean* 16:403–423.

Garrod, D. J. 1982. *Stock and recruitment—again.* MAFF, Fisheries Research Technical Report 68. 22 pp.

Garstang, W. 1899. On the variation, races, and migrations of the mackerel *(Scomber scombrus). Journal of the Marine Biological Association U.K.* 5:235–295.

Garstang, W. 1929. *The origin and evolution of larval forms*. British Association for the Advancement of Science, Report of 96th meeting, pp. 77–98.

George, D. G. and G. P. Harris. 1985. The effect of climate on long-term changes in the crustacean zooplankton biomass of Lake Windermere, U.K. *Nature* 316:536–539.

Gerritsen, J. 1980. Sex and parthenogenesis. *American Naturalist* 115:718–742.

Ghiselen, M. T. 1969. *The Triumph of the Darwinian Method*. University of California Press, Berkeley. 287 pp.

Gliwicz, M. Z. 1986. Predation and evolution of vertical migration in zooplankton. *Nature* 320:746–748.

Goldschmidt, R. 1940. *The Material Basis of Evolution*. Yale University Press, New Haven. 436 pp.

Gould, S. J. 1980. G. G. Simpson, paleontology and the modern synthesis. In *The Evolutionary Synthesis*, ed. E. Mayr and W. B. Provine (Harvard University Press, Cambridge), pp. 153–172.

———. 1985. The paradox of the first tier: An agenda for paleobiology. *Paleobiology* 11:2–12.

Gould, S. J. and R. C. Lewontin. 1979. The spandrals of San Marco and the Panglossian paradigm: A critique of the adaptationist programme. *Proceedings of the Royal Society of London B Biological Sciences* 205:581–598.

Graham, J. J. 1972. Retention of larval herring within the Sheepscot Estuary of Maine. *Fisheries Bulletin* (U.S.) 70:299–305.

———. 1982. Production of larval herring, *Clupea harengus*, along the Maine coast, 1964–1978. *Journal of Northwest Atlantic Fishery Science* 3:63–85.

Grainger, R. J. R. 1980. The distribution and abundance of early herring *(Clupea harengus* L.) larvae in Galway Bay in relation to oceanographic conditions. *Proceedings of the Royal Irish Academy Section B Biological, Geological and Chemical Science* 80B:47–60.

Gran, H. H. and T. Braarud. 1935. A quantitative study of the phytoplankton in the Bay of Fundy and the Gulf of Maine (including observations on hydrography, chemistry, and turbidity). *Journal of the Biological Board of Canada* 1:279–433.

Greenberg, D. A. 1983. Modeling the mean barotropic circulation in the Bay of Fundy and Gulf of Maine. *Journal of Physical Oceanography* 13:886–904.

Gulick, J. G. 1890. Divergent evolution through cumulative segregation. *Journal of the Linnaean Society (Zool.)* 20:215.

Haist, V. and M. Stocker. 1985. Growth and maturation of Pacific herring *(Clupea hargenus pallasi)* in the Strait of Georgia. *Canadian Journal of Fisheries and Aquatic Sciences* 42(Suppl. 1):138–146.

Halliday, R. G. and F. D. McCracken. 1970. Movements of haddock tagged off Digby, Nova Scotia. *International Commission of Northwest Atlantic Fisheries Research Bulletin* 7:8–14.

Hamre, J. 1980. Biology, exploitation, and management of the northeast Atlantic mackerel. *Rapports et Procès-verbaux des Réunions, Conseil international pour l'Exploration de la Mer* 177:212–242.

Hansen, B., D. Ellett, and D. Meldrum. 1986. *Evidence for an anticyclonic circulation on Faroe Bank*. International Council for the Explortion of the Sea, CM 1986/C:15.

Harden Jones, F. R. 1968. *Fish Migration*. Edward Arnold Ltd., London. 325 pp.

Harding, E. F. 1985. On the homogeneity of the European eel population *(Anguilla anguilla)*. *Dana* 4:49–66.

Hardy, A. C. 1953. Some problems of pelagic life. In *Essays in Marine Biology: The Richard Elmhirst Memorial Lectures* (Oliver and Boyd, Edinburgh), pp. 101–121.

———. 1954. Escape from specialization. In *Evolution as a Process*, ed. J. Huxley, A. C. Hardy, and E. B. Ford (George Allen & Unwin, London), pp. 122–142.

———. 1967. *Great Waters*. Collins, London. 542 pp.

Hardy, A. C. and E. R. Gunther. 1935. The plankton of South Georgia whaling grounds and adjacent waters, 1926–27. *Discovery Reports* 11:1–466.

Harris, J. E. 1963. The role of endogenous rhythms in vertical migration. *Journal of the Marine Biological Association U.K.* 43:153–166.

Hasler, A. D., A. T. Scholz, and R. M. Horrall. 1978. Olfactory imprinting and homing in salmon. *American Science* 66:347–355.

Hedgecock, D. 1986. Is gene flow from pelagic larval dispersal important in the adaptation and evolution of marine invertebrates? *Bulletin of Marine Science* 39:550–564.

Heincke, F. 1878. Die Varietäten des Herings I. *Jahresbuch, Kommission für die Untersuchungen der Deutschen Meere in Kiel* 4–6:37–132.

———. 1882. Die Varietäten des Herings II. *Jahresbuch, Kommission für die Untersuchungen der Deutschen Meere in Kiel* 7–11:1–86.

———. 1898. *Naturgeschichte des Herings I. Die Lokalformen und die Wanderungen des Herings in den europäischen Meeren.* Abhandlungen der deutschen Seefischereivereins, vol. 2. O. Salle, Berlin.

Helland-Hansen, B. and F. Nansen. 1909. *The Norwegian Sea.* Reports of the Norwegian Fisheries and Marine Investigations, vol. 2, pt. 1.

Henri, M., J. J. Dodson, and H. Powles. 1985. Spatial configurations of young herring *(Clupea harengus harengus)* larvae in the St. Lawrence Estuary: Importance of biological and physical factors. *Canadian Journal of Fisheries and Aquatic Sciences* 42:(Suppl. 1):91–104.

Hislop, J. R. G. 1984. A comparison of the reproductive tactics and strategies of cod, haddock, whiting, and Norway pout in the North Sea. In *Fish Reproduction: Strategies and Tactics*, ed. G. W. Potts and R. J. Wooton (Academic Press, London), pp. 311–329.

Hjort, J. 1914. Fluctuations in the great fisheries of northern Europe. *Rapports et Procès-verbaux des Réunions, Conseil international pour l'Exploration de la Mer* 20:1–228.

———. 1926. Fluctuations in the year classes of important food fishes. *Journal du Conseil* 1:1–38.

Hjort, J. 1934. *The restrictive law of population.* Huxley Memorial Lecture, Imperial College of Science and Technology, London. 46 pp.

Holton, G. 1978. *The Scientific Imagination: Case Studies.* Cambridge University Press, Cambridge. 382 pp.

Hull, D. L. 1984. Lamarck among the Anglos. In *Zoological Philosophy*, ed. J. B. Lamarck, trans. H. Elliot (University of Chicago Press, Chicago), pp. xi–lxvi.

Hulsemann, K. 1985. Two species of *Drepanopus* Brady (Copepoda:Calanoida) with discrete ranges in the Southern Hemisphere. *Journal of Plankton Research* 7:909–925.

Huntley, M. 1981. Nonselective, nonsaturated feeding by three calanid copepod species in the Labrador Sea. *Limnology and Oceanography* 26:831–842.

Huntley, M. and C. Boyd. 1984. Food-limited growth of marine zooplankton. *American Naturalist* 124:455–478.

Hutchinson, G. E. 1978. *An Introduction to Population Ecology.* Yale University Press, New Haven. 260 pp.

Huxley, T. H. 1881. The herring. *Nature* 23:607–613.

Iles, T. D. 1973. Interaction of environment and parent stock size in determining recruitment in the Pacific sardine as revealed by analysis of density-dependent O-group growth. *Rapports et Procès-verbaux des Réunions, Conseil international pour l'Exploration de la Mer* 164:228–240.

Iles, T. D. and M. Sinclair. 1982. Atlantic herring: Stock discreteness and abundance. *Science* (Washington, D.C.) 215:627–633.

———. 1985. *An instance of herring larval retention in the North Sea.* International Council for the Exploration of the Sea, C.M. 1985/H:43.

Inman, D. L. and B. M. Brush. 1973. The coastal challenge. *Science* (Washington, D.C.) 181:20–32.

International Council for the Exploration of the Sea. 1979. *The biology, distribution, and state of exploitation of fish stocks in the ICES area.* ICES Cooperative Research Report 86.

Isaacs, J. D., S. A. Tont, and G. L. Wick. 1974. Deep scattering layers: Vertical migration as a tactic for finding food. *Deep-Sea Research* 21:651–656.

Jablonski, D. and R. A. Lutz. 1983. Larval ecology of marine benthic invertebrates: Paleobiological implications. *Biological Reviews* 58:21–89.

Jablonski, D., J. J. Sepkoski, Jr., D. J. Bottjer, and P. M. Sheehan. 1983. Onshore-offshore patterns in the evolution of Phanerozoic shelf communities. *Science* (Washington, D.C.) 222:1123–1125.

Jacobs, J. 1968. Animal behavior and water movement as co-determinants of plankton distribution in a tidal system. *Sarsia* 34:355–370.

Johannes, R. E. 1978. Reproductive strategies of coastal marine fishes in the tropics. *Environmental Biology of Fishes* 3:65–84.

Johnson, D. R., B. S. Hester, and J. R. McConaugha. 1984. Studies of a wind mechanism influencing the recruitment of blue crabs in the Middle Atlantic Bight. *Continental Shelf Research* 3:425–437.

Johnson, M. W. 1960. Production and distribution of larvae of the spiny lobster *Panulirus interruptus* (Randall) with records on *P. gracilis* Streets. *Bulletin of the Scripps Institution of Oceanography* 7:413–446.

Johnson, P. O. 1977. *A review of spawning in the north Atlantic mackerel,* Scomber scombrus *L.* Fisheries Research Technical Report 27, MAFF Directorate of Fisheries Research, Lowestoft. 22 pp.

Jones, E. C. 1965. Evidence of isolation between populations of *Candacia pachydactyla* (Dana) (Copepoda:Calanoida) in the Atlantic and the Indo-Pacific Oceans. In *Proceedings, Symposium on Crustacea,* Marine Biological Association of India, Symposium Series 2, part 1, pp. 406–410.

Jones, L. T. 1969. Continuous plankton records: Studies on the zooplankton east of Newfoundland and Labrador, with particular reference to the euphausiid *Thysanoessa longicaudata* (Kroyer). *Bulletin of Marine Ecology* 6:275–300.

Jones, R. 1973. Density-dependent regulation of the numbers of cod and haddock. *Rapports et Procès-verbaux des Réunions, Conseil international pour l'Exploration de la Mer* 164:156–173.

Karnella, C. and R. H. Gibbs, Jr. 1977. The lanternfish *Lobianchia dofleini:* An example of the importance of life-history information in prediction of oceanic sound scattering. In *Oceanic Sound Scattering Prediction,* ed. N. R. Anderson and B. S. Zahurance (Plenum Press, New York), pp. 361–379.

Kean, J. 1982. Laboratory studies designed to investigate some aspects of predation on fish larvae. M.Sc. thesis, Dalhousie University, Halifax, Nova Scotia. 157 pp.

Kendall, A. W., Jr. and J. W. Reintjes. 1975. Geographic and hydrographic distribution of Atlantic menhaden eggs and larvae along the middle Atlantic coast from RV *Dolphin* cruises, 1965–66. *Fisheries Bulletin* (U.S.) 73:317–330.

Kennedy, V. S. 1982. *Estuarine Comparisons.* Academic Press, New York. 709 pp.

Kingsland, S. E. 1985. *Modeling Nature.* University of Chicago Press, Chicago.

Kleckner, R. C. and J. D. McCleave. 1985. Spatial and temporal distribution of American eel larvae in relation to north Atlantic Ocean current systems. *Dana* 4:67–92.

Kling, S. A. 1976. Relation of radiolarian distributions to subsurface hydrography in the north Pacific. *Deep-Sea Research* 23:1043–1058.

Koslow, J. A. 1984. Recruitment patterns in northwest Atlantic fish stocks. *Canadian Journal of Fisheries and Aquatic Sciences* 41:1722–1729.

Koslow, J. A. and A. Ota. 1981. The ecology of vertical migration in three common zooplankters in the La Jolla Bight, April–August 1967. *Biological Oceanography* 1:107–134.

Koslow, J. A., S. Brault, J. Dugas, and F. Page. 1985. Anatomy of an apparent year-class failure: The early life history of the 1983 Browns Bank haddock *Melanogrammus aeglefinus. Transactions of the American Fisheries Society* 114:478–489.

Koslow, J. A., K. R. Thompson, and W. Silvert. 1987. Recruitment to northwest Atlantic cod *(Gadus morhua)* and haddock *(Melanogrammus aeglefinus)* stocks: Influence of stock size and climate. *Canadian Journal of Fisheries and Aquatic Sciences* 44:26–39.

Kropach, C. 1975. The yellow-bellied sea snake, *Pelamis*, in the eastern Pacific. In *The Biology of Sea Snakes*, ed. W. A. Dunson (University Park Press, Baltimore), pp. 185–213.

Kulka, D. W., S. Corey, and T. D. Iles. 1982. Community structure and biomass of euphausiids in the Bay of Fundy. *Canadian Journal of Fisheries and Aquatic Sciences* 39:326–334.

Kyle, H. M. 1900. Contributions towards a natural history of the plaice *(P. platessa* L.). *Fisheries Board of Scotland Annual Reports* 18:189–241.

Lambert, J.-D. 1983. Contribution à l'étude des communautés planctoniques de la Baie des Chaleurs. M.Sc. thesis, Université du Québec à Rimouski. 131 pp.

Landry, M. R. 1978. Population dynamics and production of a planktonic marine copepod, *Acartia clausii*, in a small temperate lagoon in San Juan Island, Washington. *Internationale Revue der gesamten Hydrobiologie* 63:77–119.

Lasker, R. 1975. Field criteria for survival of anchovy larvae: The relation between inshore chlorophyll maximum layers and successful first feeding. *Fisheries Bulletin* (U.S.) 73:453–462.

————. 1978. The relationship between oceanographic conditions and larval anchovy food in the California Current: Identification of factors contributing to recruitment failure. *Rapports et Procès-verbaux des Réunions, Conseil international pour l'Exploration de la Mer* 173:212–230.

Laurence, G. C. 1977. A bioenergetic model for the analysis of feeding and survival potential of winter flounder, *Pseudopleuronectes americanus*, larvae during the period from hatching to metamorphosis. *Fisheries Bulletin* (U.S.) 75:529–546.

Lee, C. M. 1971. Population structures of the planktonic copepods *Centropages typicus* and *Temora longicornis* in the north Atlantic area. Ph.D. thesis, Edinburgh University.

————. 1972. Structure and function of the spermatophore and its coupling device in the centropagidae (Copepoda:Calanoida). *Bulletin of Marine Ecology* 8:1–20.

Leggett, W. C. and R. R. Whitney. 1972. Water temperature and the migration of American shad. *Fisheries Bulletin* (U.S.) 70:659–670.

Leggett, W. C., K. T. Frank, and J. E. Carscadden. 1984. Meteorological and hydrographic regulation of year-class strength in capelin *(Mallotus villosus)*. *Canadian Journal of Fisheries and Aquatic Sciences* 41:1193–1201.

Leis, J. 1982. Nearshore distributional gradients of larval fish (15 taxa) and planktonic crustaceans (6 taxa) in Hawaii. *Marine Biology* 72:89–97.

Lesser, J. H. R. 1978. Phyllosoma larvae of *Jasus edwardsii* (Hutton) (Crustacea:Decapoda:Palinuridae) and their distribution off the east coast of North Island, New Zealand. *New Zealand Journal of Marine and Freshwater Research* 12:357–370.

Lett, P. 1978. *A comparative study of the recruitment mechanisms of cod and mackerel, their interaction, and its implication for dual stock management.* Canadian Technical Report of Fisheries and Aquatic Sciences 988. 45 pp.

Levinton, J. S. 1982. *Marine Ecology.* Prentice Hall, Englewood Cliffs, N.J. 526 pp.

Lewontin, R. C. 1980. Theoretical population genetics in the evolutionary synthesis. In *The Evolutionary Synthesis*, ed. E. Mayr and W. B. Provine (Harvard University Press, Cambridge), pp. 59–68.

Lindley, J. A. 1980. Population dynamics and production of euphausiids. 2. *Thysanoessa inermis* and *T. raschii* in the North Sea and American coastal waters. *Marine Biology* 59:225–233.

Lipps, J. H. 1970. Plankton evolution. *Evolution* 24:1–22.

Lobel, P. 1978. Diel, lunar, and seasonal periodicty in the reproductive behavior of some pomacanthid fish, *Centropyge potteri* and some other reef fishes in Hawaii. *Pacific Science* 32:193–207.

Longhurst, A. R. 1976. Vertical migration. In *Ecology of the Seas*, ed. D. H. Cushing and J. J. Walsh (Blackwell Scientific Publications, Oxford), pp. 116–137.

Longhurst, A., D. Sameoto, and A. Herman. 1984. Vertical distribution of Arctic zooplankton in summer: Eastern and Canadian archipelago. *Journal of Plankton Research* 6:137–168.

Lough, R. G. 1976. Larval dynamics of the dungeness carb, *Cancer magister*, off the central Oregon coast, 1970–71. *Fisheries Bulletin* (U.S.):74:353–375.

Lux, F. E. 1963. Identification of New England yellowtail flounder groups. *Fisheries Bulletin* (U.S.): 63:1–10.

MacCall, A. D. 1984. *Population models of habitat selection, with application to the northern anchovy.* NMFS Southwest Fisheries Center, Administrative Report LJ-84-01. 98 pp.

McCleave, J. D., R. C. Kleckner, and M. Castonguay. In press. Reproductive sympatry of American and European eels and implications for migration and taxonomy. In *Common Strategies of Anadromous and Catadromous Fishes* (American Fisheries Society Symposia, Boston).

McCracken, F. D. 1959. Studies of haddock in the Passamaquoddy Bay region. *Journal of the Fisheries Research Board of Canada* 17:175–180.

McGowan, J. A. 1963. Geographical variation of the planktonic mollusc *Limacina helicina* in the north Pacific. In *Speciation in the Sea*, Systematics Assoc. Publ. 5, British Museum of Natural History, pp. 109–128.

McGowan, J. A. and E. Brinton. 1985. Martin W. Johnson (1893–1984). *Bulletin of Marine Science* 37(2):407–410.

McIntosh, R. P. 1985. *The Background of Ecology.* Cambridge University Press, Cambridge. 383 pp.

Mackintosh, N. A. 1937. The seasonal circulation of the Antarctic macroplankton. *Discovery Reports* 16:365–412.

Mclaren, I. A. 1963. Effects of temperature on growth of zooplankton and the adaptive value of vertical migration. *Journal of the Fisheries Research Board of Canada* 20:685–727.

―――. 1978. Generation lengths of some temperate marine copepods: Estimation, prediction, and implications. *Journal of the Fisheries Research Board of Canada* 35:1330–1342.

Makarov, R. R. 1969. Transport and distribution of decapod larvae in the plankton of the Western Kamchatka Shelf. *Oceanology* 9:251–261.

Mani, G. S. 1982. Genetic diversity and ecological stability. In *Evolutionary Ecology*, ed. B. Shorrocks (Blackwell Scientific Publications, Oxford), pp. 363–397.

Margalef, R. 1978. General concepts of population dynamics and food links. In *Marine Ecology*, ed. O. Kinne (Wiley Interscience, New York), vol. 4, pp. 617–704.

―――. 1985. From hydrodynamic processes to structure (information) and from information to process. In *Ecosystem Theory for Biological Oceanography*, ed. R. E. Ulanowicz and T. Platt, *Canadian Bulletin of Fisheries and Aquatic Sciences* 213:200–220.

Marliave, J. B. 1981. Vertical migrations and larval settlement in *Gilbertidia sigalutes*, F. Cottidae. *Rapports et Procès-verbaux des Réunions, Conseil international pour l'Exploration de la Mer* 178:349–351.

―――. 1986. Lack of dispersal of rocky intertidal fish larvae. *Transactions of the American Fisheries Society* 115:149–154.

Marshall, N. B. 1979. *Developments in Deep-Sea Biology.* Blandford Press, Poole. 566 pp.

Masterman, A. T. 1985. On the rate of growth of the food fishes. *Fishery Board of Scotland Annual Reports* 14:294–302.

Mauchline, J. 1960. The biology of the euphausiid crustacean, *Meganyctiphanes norvegica* (M. Sars). *Proceedings of the Royal Society of Edinburgh B* (Biology) 67:141–179.

―――. 1965. The larval development of the euphausiid, *Thysanoessa* (M. Sars). *Crustaceana* 9:31–40.

―――. 1966. The biology of *Thysanoessa raschii* (M. Sars), with a comparison of its diet with that of *Meganyctiphanes norvegica* (M. Sars). In *Some Contemporary Studies in Marine Science*, ed. H. Barnes (George Allen & Unwin, London), pp. 493–510.

————. 1985. Growth and production of Euphausiacea (Crustacea) in the Rockall Trough. *Marine Biology* 90:19–26.

Mauchline, J. and L. R. Fisher. 1969. The biology of euphausiids. *Advances in Marine Biology* 7:1–454.

Mayr, E. 1942. *Systematics and the Origin of Species*. Columbia University Press, New York. 334 pp.

————. 1954. Change of genetic environment and evolution. In *Evolution as a Process*, ed. J. Huxley, A. C. Hardy, and E. B. Ford (Allen & Unwin, London), pp. 157–180.

————. 1957a. Species concepts and definitions. In *The Species Problem*, ed. E. Mayr, A.A.A.S. Publ. 50, Washington, D.C., pp. 1–22.

————. 1957b. Difficulties and importance of the biological species concept. In *The Species Problem*, ed. E. Mayr, A.A.A.S. Publ. 50, Washington, D.C., pp. 371–388.

————. 1961. Cause and effect in biology. *Science* 134:1501–1506.

————. 1980. Prologue: Some thoughts on the history of the evolutionary synthesis. In *The Evolutionary Synthesis*, ed. E. Mayr and W. B. Provine (Harvard University Press, Cambridge), pp. 1–48.

————. 1982a. *The Growth of Biological Thought*. Harvard University Press, Cambridge. 974 pp.

————. 1982b. Speciation and macroevolution. *Evolution* 36:119–132.

Mayr, E. and W. B. Provine, eds. 1980. *The Evolutionary Synthesis*. Harvard University Press, Cambridge. 487 pp.

Melvin, G. D., J. D. Michael, and J. D. Martin. 1985. Fidelity of American shad *Alosa sapidissima* (Osteichthyes:Clupeidae) to its river of previous spawning. *Canadian Journal of Fisheries and Aquatic Sciences* 42:1–19.

Methot, R. D., Jr. 1981. Growth rates and age distributions of larval and juvenile northern anchovy, *Engraulis mordax*, with inferences on larval survival. Ph.D. thesis, University of California, San Diego. 200 pp.

————. 1983. Seasonal variation in survival of larval northern anchovy, *Engraulis mordax*, estimated from the age distribution of juveniles. *Fisheries Bulletin* (U.S.):81:741–750.

Meyer, H. A. 1878. Beobachtungen über das Wachsthum des Herings im westlichen Theile der Ostsee. In *Jahresbericht der Kommission zur wissenschaftlichen Untersuchung der deutschen Meere in*

Kiel für die Jahre 1874, 1875, 1876, ed. H. A. Meyer, M. Möbius, G. Karsten, and V. Hensen (Wiegandt, Hempel and Parey), pp. 227–252.

Miller, C. B. 1983. The zooplankton of estuaries. In *Estuaries and Enclosed Seas,* ed. B. H. Ketchum (Elsevier, Amsterdam), pp. 103–149.

Miller, D., J. B. Colton, Jr., and R. R. Marak. 1963. A study of the vertical distribution of larval haddock. *Journal du Conseil* 28:37–49.

Mivart, St. G. 1871. *The Genesis of Species.* Appleton, New York.

Murray, B. G. 1967. Dispersal in vertebrates. *Ecology* 48:975–978.

Needham, J. 1930. On the penetration of marine organisms into freshwater. *Biologisches Zentralblatt* 50:504–509.

Nelson, W. R., M. Ingham, and W. E. Schaaf. 1977. Larval transport and year class strength of Atlantic menhaden, *Brevoortia tyrannus. Fisheries Bulletin* (U.S.) 75:23–41.

Newton, I. and M. Marquiss. 1983. Dispersal of sparrowhawks between birthplace and breeding place. *Journal of Animal Ecology* 52:463–477.

Nichols, J. H., B. M. Thompson, and M. Cryer. 1982. Production, drift, and mortality of the planktonic larvae of the edible crab (*Cancer pagurus*) off the north-east coast of England. *Netherlands Journal of Sea Research* 16:173–184.

Nihoul, J. J. and F. C. Ronday. 1975. The influence of the "tidal stress" on the residual circulation: Application to the Southern Bight of the North Sea. *Tellus* 27:484–489.

Norcross, B. L., H. M. Austin, and S. K. LeDuc. 1984. *A statistical model of climatic factors affecting recruitment of Atlantic croaker* (Micropogonias undulatus). International Council for the Exploration of the Sea, C.M. 1984/D:11.

O'Boyle, R. N., M. Sinclair, R. J. Conover, K. H. Mann, and A. C. Kohler. 1984. Temporal and spatial distribution of ichthyoplankton communities of the Scotian Shelf in relation to biological, hydrological, and physiographic features. *Rapports et Procès-verbaux des Réunions, Conseil international pour l'Exploration de la Mer* 183:27–40.

O'Dor, R. K. 1981. *Illex illecebrosus.* In *Cephalopod Life Cycles,* ed. P. R. Boyle (Academic Press, London), pp. 175–199.

Ohman, M. D. 1985. Resource-satiated population growth of the copepod *Pseudocalanus* sp. *Archiv für Hydrobiologie* 21:15–32.

Ohman, M. D., B. W. Frost, and E. B. Cohen. 1983. Reverse diel vertical migration: An escape from invertebrate predators. *Science* 220:1404–1407.

Ottestad, P. 1932. On the biology of some southern copepoda. *Hvalradets Skrifter* 5:1–61.

Ouellet, P. and J. J. Dodson. 1985. Dispersion and retention of anadromous rainbow smelt *(Osmerus mordax)* larvae in the 201 middle estuary of the St. Lawrence River. *Canadian Journal of Fisheries and Aquatic Sciences* 42:332–341.

Owen, G. 1984. Seasonal variation in the composition and abundance of near shore ichthyoplankton off northwest Puerto Rico. Abstract. *Eos* 65(45):928.

Palmer, A. R. 1985. Quantum changes in gastropod shell morphology need not reflect speciation. *Evolution* 39:699–705.

Palmer, A. R. and R. R. Strathmann. 1981. Scale of dispersal in varying environments and its implications for life histories of marine invertebrates. *Oecologia* 48:308–318.

Parrish, R. H., C. S. Nelson, and A. Bakun. 1981. Transport mechanisms and reproductive success of fishes in the California Current. *Biological Oceanography* 1:175–203.

Paterson, H. E. H. 1982. Perspectives on speciation by reinforcement. *South African Journal of Science* 78:53–57.

Pearcy, W. G. 1962. Ecology of an estuarine population of winter flounder, *Pseudopleuronectes americanus* (Walbaum). II. Distribution and dynamics of larvae. *Bulletin of the Bingham Oceanographic Collection* 18:16–38.

Pearson, K. 1906. Walter Frank Raphael Weldon. *Biometrika* 5:1–52.

Pepin, P. 1985. The influence of variations of abundance on the foraging dynamics of Atlantic mackerel, *Scomber scombrus*, and its importance in modulating the impact of adult pelagic fish on larval fish survival and recruitment variability. Ph.D. thesis, Dalhousie University, Halifax, Nova Scotia.

Perlmutter, A. 1947. The blackback flounder and its fishery in New England and New York. *Bulletin of the Bingham Oceanographic Collection* 11:1–92.

Peterson, W. T. 1985. Abundance, age structure and in situ egg production rates of the copepod *Temora longicornis* in Long Island Sound, New York. *Bulletin of Marine Science* 37:726–738.

Peterson, W. T., C. B. Miller, and A. Hutchinson. 1979. Zonation and maintenance of copepod populations in the Oregon upwelling zone. *Deep-Sea Research* 26:467–494.

Petit, D. and C. Courties. 1976. *Calanoides carinatus* (copépode pélagique) sur le plateau continental. Congolais 1. Aperçu sur la répartition bathymétrique, géographique et biométrique des stades; générations durant la saison froide 1974. *Cahiers O.R.S.T.O.M.*, *Série Océanographie* 14:177–199.

Phillips, B. P. 1981. The circulation of the southeastern Indian Ocean and the planktonic life of the western rock lobster. *Oceanography and Marine Biology, An Annual Review* 19:11–39.

Pianka, E. R. 1979. *Evolutionary Ecology*, 2d ed. Harper and Row, New York. 397 pp.

Pingree, R. D. and D. K. Griffiths. 1978. Tidal fronts on the shelf seas around the British Isles. *Journal of Geological Research* 83:4615–4622.

———. 1980. A numerical model of the M2 tide in the Gulf of St. Lawrence. *Oceanologica Acta* 3:221–225.

Pingree, R. D. and L. Maddock. 1985. Stokes, Euler and Lagrange aspects of residual tidal transports in the English Channel and the Southern Bight of the North Sea. *Journal of the Marine Biological Association U.K.* 65:969–982.

Platt, T. 1985. Structure of the marine ecosystem: Its allometric basis. In *Ecosystem Theory for Biological Oceanography*, ed. R. E. Ulanowicz and T. Platt. *Canadian Bulletin of Fisheries and Aquatic Sciences* 213:55–64.

Platt, T. and W. G. Harrison. 1985. Biogenic fluxes of carbon and oxygen in the ocean. *Nature* 318:55–58.

Popova, V. P. 1972. Characteristics of the reproductive biology of the Black Sea turbot *(Scophthalmus maeoticus maeoticus* [Pallas]) (observations in the sea). *Journal of Ichthyology* 12:961–967.

Pringle, J. D. 1986. California spiny lobster *(Panulirus interruptus)* larval retention and recruitment: A review and synthesis. *Canadian Journal of Fisheries and Aquatic Sciences* 43:2142–2152.

Redfield, A. C. 1939. The history of a population of *Limacina retroversa* during its drift across the Gulf of Maine. *Biological Bulletin* (Woods Hole) 76:26–27.

Redfield, A. C. and A. C. Beale. 1940. Factors determining the distribution of populations of chaetognaths in the Gulf of Maine. *Biological Bulletin* (Woods Hole) 79:459–487.

Rensch, B. 1929. *Das Prinzip geographischer Rassenkreise und das Problem der Artbildung.* Borntraeger, Berlin.

———. 1959. *Evolution Above the Species Level.* John Wiley, New York. 419 pp.

Rice, A. L. and I. Kristensen. 1982. Surface swarms of swimming crab megalopae at Curaçao (Decapoda, Brachyura). *Crustaceana* 42:233–240.

Ricker, W. E. 1954. Stock and recruitment. *Journal of the Fisheries Research Board of Canada* 11:559–623.

Riley, G. A. 1946. Factors controlling phytoplankton populations on Georges Bank. *Journal of Marine Research* 6:49–68.

Ringo, J., D. Wood, R. Rockwell, and H. Dowse. 1985. An experiment testing two hypotheses of speciation. *American Naturalist* 126:642–661.

Robertson, D. R., S. G. Hoffman, and J. Sheldon. 1981. Availability of space for the Caribbean damselfish *Eupomacentrus planifrons.* *Ecology* 62:1162–1169.

Rogers, H. 1940. Occurrence and retention of plankton within an estuary. *Journal of the Fisheries Research Board of Canada* 5:164–171.

Ronday, F. C. 1975. Mesoscale effects of the "tidal stress" on the residual circulation of the North Sea. *Mémoirs de la Société Royale des Sciences de Liège (6e Série)* 7:273–287.

Rose, M. 1925. Contribution à l'étude de la biologie du plankton, le problème des migrations verticales journalières. *Archives de Zoologie Expérimentale et Générale* 64:387–542.

Rosenberg, A. A. 1984. Causal analysis of catch and recruitment variations in exploited fish populations. Ph.D. thesis, Dalhousie University, Halifax, Nova Scotia. 224 pp.

Rothlisburg, P. A. and C. B. Miller. 1983. Factors affecting the distribution, abundance, and survival of *Pandalus jordani* (Decapoda, Pandalidae) larvae off the Oregon coast. *Fisheries Bulletin* (U.S.) 81:455–472.

Rowell, T. W., R. W. Trites, and E. C. Dawe. 1984. *Larval and juvenile distribution of the short-finned squid* (Illex illecebrosus) *in rela-*

tion to the Gulf Stream frontal in the Blake Plateau and Cape Hatteras area. Northwest Atlantic Fisheries Organization SCR Doc. 84/IX/111. 37 pp.

Runge, J. A. 1981. Egg production of *Calanus pacificus* Brodsky and its relationship to seasonal changes in phytoplankton availability. Ph.D. thesis, University of Washington, Seattle.

Russell, F. S. 1927. The vertical distribution of plankton in the sea. *Biological Reviews of the Cambridge Philosophical Society* 2:213–262.

Saila, S. B. 1961. A study of winter flounder movements. *Limnology and Oceanography* 6:292–298.

Sale, P. F. 1970. Distribution of larval Acanthuridae off Hawaii. *Copeia* 1970: 765–766.

———. 1980. The ecology of fishes on coral reefs. *Oceanography and Marine Biology, an Annual Review* 18:367–421.

Sandifer, P. A. 1973. Distribution and abundance of decapod crustacean larvae in the York River estuary and adjacent lower Chesapeake Bay, Virginia, 1968–1969. *Chesapeake Science* 14:235–237.

———. 1975. The role of pelagic larvae in recruitment to populations of adult decapod crustaceans in the York River estuary and adjacent lower Chesapeake Bay, Virginia. *Estuarine and Coastal Marine Science* 3:269–279.

Saville, A. 1956. Eggs and larvae of haddock (*Gadus aeglefinus* L.) at Faroe. *Marine Research of Scotland* 4:1–27.

———, ed. 1980. The assessment and management of pelagic fish stocks. *Rapports et Procès-verbaux des Réunions, Conseil international pour l'Exploration de la Mer* 177:1–517.

Scheltema, R. S. 1974. Biological interactions determining larval settlement of marine invertebrates. *Thallassia Jugoslavica* 10:263–296.

Schmaus, P. H. 1917. Die Rhincalanus-Arten, ihre Systematik, Entwicklung und Verbreitung. *Zoologischer Anzeiger* 48:305–319.

Schmaus, P. H. and K. Lehnhofer. 1927. Systematik und Verbreitung der Gattung copepoda 4: *Rhincalanus* Dana 1852. In *Wissenschaftliche Ergebnisse der Deutschen Tiefsee-Expedition auf dem Dampfer "Valdivia" 1898–1899* (Gustav Fisher, Jena), vol. 23, no. 8, pp. 358–399.

Schmidt, J. 1909. The distribution of the pelagic fry and the spawning regions of the gadoids in the north Atlantic from Iceland to Spain. *Rapports et Procès-verbaux des Réunions, Conseil international pour l'Exploration de la Mer* 10(4):1–229.

———. 1917. Racial investigations. I. *Zoarces viviparus* L. and local races of the same. *Comptes-rendus des Travaux du Laboratoire Carlsberg* 13(3):279–396.

———. 1922. The breeding places of the eel. *Philosophical Transactions of the Royal Society of London B* 211:179–208.

———. 1930. Racial investigations. X. The Atlantic cod *(Gadus callarias* L.) and local races of the same. *Comptes-rendus des Travaux du Laboratoire Carlsberg* 18(6):1–71.

Scott, J. S. 1982. Distribution of juvenile haddock around Sable Island on the Scotian Shelf. *Journal of Northwest Atlantic Fishery Science* 3:87–90.

———. 1984. Short-term changes in distribution, size and availability of juvenile haddock around Sable Island off Nova Scotia. *Journal of Northwest Atlantic Fishery Science* 5:109–112.

Sepkoski, J. 1987. Environmental trends in extinction during the Paleozoic. *Science* 235:64–66.

Sette, O. E. 1943. Biology of the Atlantic mackerel *(Scomber scombrus)* of North America. *Fisheries Bulletin* (U.S.) 50:149–237.

Sheldon, R. W., A. Prakash, and W. H. Sutcliffe, Jr. 1972. The size distribution of particles in the ocean. *Limnology and Oceanography* 17:327–340.

Sheldon, R. W., W. H. Sutcliffe, Jr., and A. Prakash. 1973. The production of particles in the surface waters of the ocean with particular reference to the Sargasso Sea. *Limnology and Oceanography* 18:719–733.

Sheldon, R. W., W. H. Sutcliffe, Jr., and M. A. Paranjape. 1977. Structure of pelagic food chain and relationship between plankton and fish production. *Journal of the Fisheries Research Board of Canada* 34:2344–2353.

Shepherd, J. G. 1982. A versatile new stock-recruitment relationship for fisheries, and the construction of sustainable yield curves. *Journal du Conseil* 40:67–75.

Shields, W. M. 1982. *Philopatry, Inbreeding, and the Evolution of Sex.* State University of New York Press, Albany.

————. 1983. Optimal inbreeding and the evolution of philopatry. In *The Ecology of Animal Movement*, ed. I. R. Swingland and P. J. Greenwood (Clarendon Press, Oxford), pp. 132–159.

Simpson, G. G. 1944. *Tempo and Mode in Evolution.* Columbia University Press, New York.

Simpson, H. J. and J. R. Hunter. 1974. Fronts in the Irish Sea. *Nature* 1250:404–406.

Sinclair, A., M. Sinclair, and T. D. Iles. 1981. An analysis of some biological characteristics of the 4X juvenile-herring fishery. *Proceedings of the Nova Scotia Institute of Science* 31:155–171.

Sinclair, M. 1978. Summer phytoplankton variability in the lower St. Lawrence Estuary. *Journal of the Fisheries Research Board of Canada* 35:1171–1185.

Sinclair, M. and T. D. Iles. 1985. Atlantic herring *(Clupea harengus)* distributions in the Gulf of Maine–Scotian Shelf area in relation to oceanographic features. *Canadian Journal of Fisheries and Aquatic Sciences* 42:880–887.

Sinclair, M. and M. J. Tremblay. 1984. Timing of spawning of Atlantic herring *(Clupea harengus harengus)* populations and the match-mismatch theory. *Canadian Journal of Fisheries and Aquatic Sciences* 41:1055–1065.

Sinclair, M., D. V. Subba Rao, and R. Couture. 1981. Phytoplankton temporal distributions in estuaries. *Oceanologica Acta* 4:239–246.

Sinclair, M., A. Sinclair, and T. D. Iles. 1982. Growth and maturation of southwest Nova Scotia Atlantic herring *(Clupea harengus harengus).* *Canadian Journal of Fisheries and Aquatic Sciences* 39:288–295.

Sinclair, M., J. J. Maguire, P. Koeller, and J. S. Scott. 1984. Trophic dynamic models in light of current resource inventory data and stock assessment results. *Rapports et Procès-verbaux des Réunions, Conseil international pour l'Exploration de la Mer* 183:269–284.

Sinclair, M., V. C. Anthony, T. D. Iles, and R. N. O'Boyle. 1985a. Stock assessment problems in Atlantic herring *(Clupea harengus)* in the northwest Atlantic. *Canadian Journal of Fisheries and Aquatic Sciences* 42:888–898.

Sinclair, M., R. K. Mohn, G. Robert, and D. L. Roddick. 1985b. *Considerations for the effective management of Atlantic scallops.*

Canadian Technical Report of Fisheries and Aquatic Sciences 1382. 113 pp.

Sinclair, M., M. J. Tremblay, and P. Bernal. 1985c. El Niño events and variability in a Pacific mackerel *(Scomber japonicus)* survival index: Support for Hjort's second hypothesis. *Canadian Journal of Fisheries and Aquatic Sciences* 42:602–608.

Sinclair, M., G. L. Bugden, C. L. Tang, J.-C. Therriault, and P. A. Yeats. 1986. Assessment of effects of freshwater runoff variability on fisheries production in coastal waters. In *The Role of Freshwater Outflow in Coastal Marine Ecosystems*, ed. S. Skreslet, NATO ASI Series G: Ecological Sciences 7 (Springer-Verlag, Berlin), pp. 139–160.

Sissenwine, M. P. 1985. Why do fish populations vary? In *Dahlem Workshop on Exploitation of Marine Communities*, ed. R. May (Springer-Verlag, Berlin).

Sloan, N. A. 1985. Life history characteristics of fjord-dwelling king crabs *Lithodes aequispina*. *Marine Ecology Progress Series* 22:219–228.

Slobodkin, L. B. 1972. On the inconstancy of ecological efficiency and the form of ecological theories. *Transactions of the Connecticut Academy of Arts and Sciences* 44:291–305.

Slobodkin, L. B., F. E. Smith, and N. G. Hairston. 1967. Regulation in terrestrial ecosystems, and the implied balance of nature. *American Naturalist* 101:109–124.

Smith, W. G. and W. W. Morse. 1985. Retention of larval haddock *Melanogrammus aeglefinus* in the Georges Bank region, a gyre influenced spawning area. *Marine Ecology Progress Series* 24:1–13.

Smith, W. G., J. D. Sibunka, and A. Wells. 1978. Diel movements of larval yellowtail flounder, *Limanda ferruginea*, determined from discrete depth sampling. *Fisheries Bulletin* (U.S.) 76:167–178.

Somme, I. D. 1933. A possible relation between the production of animal plankton and the current system of the sea. *American Naturalist* 67:30–52.

———. 1934. Animal plankton of the Norwegian coast waters and the open sea. *Fiskeridirektoratets Skrifter, Serie Havundersokelser* 4(9):1–163.

Southward, A. J. 1980. The Western English Channel—an inconstant ecosystem? *Nature* 285:361–366.

Spight, T. M. 1974. Sizes of populations of a marine snail. *Journal of Ecology* 55:712–729.

Stearns, S. C. 1982. The emergence of evolutionary and community ecology as experimental sciences. *Perspectives in Biology and Medicine* 25:621–648.

Stebbins, G. L. 1950. *Variation and Evolution in Plants*. Columbia University Press, New York.

Stebbins, G. L. and F. J. Ayala. 1985. The evolution of Darwinism. *Scientific American* 253(July):72–82.

Steele, J. H. 1985. A comparison of terrestrial and marine ecological systems. *Nature* 313:355–358.

Steele, J. H. and E. W. Henderson. 1984. Modeling long-term fluctuations in fish stocks. *Science* 224:985–987.

Strathmann, R. R. 1982. Selection for retention or export of larvae in estuaries. In *Estuarine Comparisons*, ed. V. S. Kennedy (Academic Press, New York), pp. 521–536.

———. 1985. Feeding and nonfeeding larval development and life-history evolution in marine invertebrates. *Annual Review of Ecology and Systematics* 16:339–361.

Sulkin, S. D. and W. Van Heukelem. 1982. Larval recruitment in the crab *Callinectes sapidus* Rathbun: An amendment to the concept of larval retention in estuaries. In *Estuarine Comparisons*, ed. V. S. Kennedy (Academic Press, New York), pp. 459–475.

Sund, O. 1924. Snow and the survival of cod fry. *Nature* 113:163.

Sutcliffe, W. H., Jr. 1972. Some relations of land drainage, nutrients, particulate material, and fish catch in two eastern Canadian bays. *Journal of the Fisheries Research Board of Canada* 29:357–362.

———. 1973. Correlations between seasonal river discharge and local landings of American lobster *(Homarus americanus)* and Atlantic halibut *(Hippoglossus hippoglossus)* in the Gulf of St. Lawrence. *Journal of the Fisheries Research Board of Canada* 30:856–859.

Sutcliffe, W. H., Jr., K. Drinkwater, and B. S. Muir. 1977. Correlations of fish catch and environmental factors in the Gulf of Maine. *Journal of the Fisheries Research Board of Canada* 34:19–30.

Sutcliffe, W. H., Jr., R. H. Loucks, K. Drinkwater, and A. R. Coote. 1983. Nutrient flux onto the Labrador Shelf from Hudson Strait and its biological consequences. *Canadian Journal of Fisheries and Aquatic Sciences* 40:1692–1701.

Sverdrup, H. U. 1953. On conditions for the vernal blooming of phytoplankton. *Journal du Conseil* 18:287–295.

Taggart, C. T. 1986. Mortality of larval capelin *(Mallotus villosus* Müller): Environmental and density correlates during post-emergent dispersal. Ph.D. thesis, McGill University, Montréal. 199 pp.

Talbot, F. H., B. C. Russell, and G. R. Anderson. 1978. Coral reef fish communities: Unstable, high-diversity systems? *Ecological Monographs* 48:425–440.

Talbot, J. W. 1974. Diffusion studies in fisheries biology. In *Sea Fisheries Research*, ed. F. R. Harden-Jones (Elek Science, London), pp. 31–54.

Templeman, W. 1962. *Divisions of cod stocks in the northwest Atlantic.* International Commission of Northwest Atlantic Fisheries, Selected Paper 79-123.

Thompson, V. 1976. Does sex accelerate evolution? *Evolutionary Theory* 1:131–156.

———. 1977. Recombination and response to selection in *Drosophila melanogaster. Genetics* 85:125–140.

Thorson, G. 1950. Reproductive and larval ecology of marine bottom invertebrates. *Biological Reviews* 25:1–45.

Thresher, R. E. and E. B. Brothers. 1985. Reproductive ecology and biogeography of Indo-West Pacific angelfishes (Pisces:Pomacanthidae). *Evolution* 39:878–887.

Trexler, J. C. 1984. Aggregation and homing in a chrysidid wasp. *Oikos* 43:133–137.

Trinast, E. M. 1975. Tidal currents and *Acartia* distribution in Newport Bay, California. *Estuarine and Coastal Marine Science* 3:165–176.

Trites, R. W. 1983. Physical oceanographic features and processes relevant to *Illex illecebrosus* spawning areas and subsequent larval distribution. *Northwest Atlantic Fisheries Organization, Scientific Council Studies* 6:34–55.

Tyler, A. V. and S. J. Westrheim. 1985. Effects of transport, temperature, and stock size on recruitment of Pacific cod *(Gadus macrocephalus)*. Paper read at 1985 Symposium of the International North Pacific Fisheries Commission, Tokyo, October 1985. Department of Fisheries and Oceans, Pacific Biological Station, Nanaimo, B.C. 18 pp.

Valentine, J. W. 1985. *Phanerozoic Diversity Patterns*. Princeton University Press, Princeton. 441 pp.

Valentine, J. W. and Jablonski. 1983. Speciation in the shallow sea: General patterns and biogeographic controls. In *Evolution, Time and Space: The Emergence of the Biosphere*, ed. R. W. Sims, J. H. Price, and P. E. S. Whalley (Academic Press, London), pp. 201–226.

Van der Spoel, S. and R. P. Heyman. 1983. *A Comparative Atlas of Zooplankton*. Springer-Verlag, Berlin. 186 pp.

Van Valen, L. 1973a. A new evolutionary law. *Evolutionary Theory* 1:1–30.

———. 1973b. Pattern and the balance of nature. *Evolutionary Theory* 1:31–49.

———. 1976. Energy and evolution. *Evolutionary Theory* 1:179–229.

———. 1982. Why misunderstand the evolutionary half of biology? In *Conceptual Issues in Ecology*, ed. E. Saarinen (D. Reidel, Dorddrecht), pp. 323–343.

Victor, B. C. 1983. Recruitment and population dynamics of a coral reef fish. *Science* 219:419–420.

———. 1986. Delayed metamorphosis with reduced larval growth in a coral reef fish *(Thalassoma bifasciatum)*. *Canadian Journal of Fisheries and Aquatic Sciences* 43:1028–1213.

Voris, H. K. 1985. Population size estimates for a marine snake *(Enhydrina schistosa)* in Malaysia. *Copeia* 1985:955–961.

Wallace, A. R. 1981. *Darwinism*. MacMillan and Co., London. 494 pp.

Ware, D. M. 1980. Bioenergetics of stock and recruitment. *Canadian Journal of Fisheries and Aquatic Sciences* 43:1028–1213.

Watson, W. and J. M. Leis. 1974. *Ichthyoplankton in Kanehoe Bay, Hawaii: A one-year study of fish eggs and larvae*. University of Hawaii Sea Grant Technical Report TR-75-01. 178 pp.

Weisman, A. 1886. *Die Bedeutung der sexuellen Fortpflanzung für die Selektionstheorie*. Gustav Fisher, Jena.

Wiborg, K. F. 1952. Fish eggs and larvae along the coast of northern Norway during April–June 1950 and 1951. *Annales Biologiques* 8:11–16.

————. 1960a. Investigations on eggs and larvae of commercial fishes in Norwegian coastal and offshore waters in 1957–58. *Fiskeridirektoratets Skrifter, Serie Havundersokelser* 12(7):1–27.

————. 1960b. Investigations on pelagic fry of cod and haddock in coastal and offshore areas of northern Norway in July–August 1957. *Fiskeridirektoratets Skrifter, Serie Havundersokelser* 12(8):1–16.

Williams, D. M. 1980. Dynamics of the Pomacentrid community on small patch reefs in one tree lagoon (Great Barrier Reef). *Bulletin of Marine Science* 30:159–170.

Williams, G. C. and R. K. Koehn. 1984. Population genetics of north Atlantic catadromous eels *(Anguilla)*. In *Evolutionary Genetics of Fishes*, ed. B. J. Turner (Plenum Press, New York), pp. 529–560.

Williams, R. and D. V. P. Conway. 1982. Population growth and vertical distribution of *Calanus helgolandicus* in the Celtic Sea. *Netherlands Journal of Sea Research* 16:185–194.

Williamson, H. C. 1900. On the mackerel on the east and west coasts of Scotland. *Reports of the Fishery Board of Scotland* 18:294–329.

Wise, J. P. 1962. Cod groups in New England area. *Fisheries Bulletin* (U.S.) 63:189–203.

Wood, L. and W. J. Hargis, Jr. 1971. Transport of bivalve larvae in a tidal estuary. In *Fourth European Marine Biology Symposium*, ed. D. J. Crisp (Cambridge University Press, London), pp. 29–44.

Wooldridge, T. and T. Erasmus. 1980. Utilization of tidal currents by estuarine zooplankton. *Estuarine and Coastal Marine Science* 11:107–114.

Wright, S. 1931. Evolution in Mendelian populations. *Genetics* 6:97–159.

————. 1934. Physiological and evolutionary theories of dominance. *American Naturalist* 68:24–53.

Wroblewski, J. S. 1982. Interaction of currents and vertical migration in maintaining *Calanus marshallae* in the Oregon upwelling zone— a simulation. *Deep-Sea Research* 29:665–686.

Wunsch, C. 1981. Low-frequency variability of the sea. In *Evolution of Physical Oceanography*, ed. B. A. Warren and C. Wunsch (Massachusetts Institute of Technology, Cambridge), pp. 342–347.

Wynne-Edwards, V. C. 1962. *Animal Dispersion in Relation to Social Behavior.* Oliver and Boyd, Edinburgh.

Zaret, T. M. and J. S. Suffern. 1976. Vertical migration in zooplankton as a predator avoidance mechanism. *Limnology and Oceanography* 21:804–813.

Zijlstra, J. J. 1970. Herring larvae in the central North Sea. *Berichte der Deutschen Wissenschaftlichen Kommission für Meeresforschung* 21:92–115.

Zinsmeister, W. J. and R. M. Feldmann. 1984. Cenozoic high latitude heterochroneity of southern hemisphere marine faunas. *Science* 224:281–283.

INDEX